Aeolian Sediments
ANCIENT AND MODERN

Aeolian Sediments

ANCIENT AND MODERN

Edited by
K. Pye and N. Lancaster

SPECIAL PUBLICATION NUMBER 16 OF THE
INTERNATIONAL ASSOCIATION OF SEDIMENTOLOGISTS
PUBLISHED BY BLACKWELL SCIENTIFIC PUBLICATIONS
OXFORD LONDON EDINBURGH BOSTON
MELBOURNE PARIS BERLIN VIENNA

and published for them by
Blackwell Scientific Publications
Editorial Offices:
Osney Mead, Oxford OX2 0EL
25 John Street, London WC1N 2BL
23 Ainslie Place, Edinburgh EH3 6AJ
238 Main Street, Cambridge
Massachusetts 02142, USA
54 University Street, Carlton
Victoria 3053, Australia

Other Editorial Offices:
Librairie Arnette SA
2, rue Casimir-Delavigne
75006 Paris
France

Blackwell Wissenschafts-Verlag GmbH
Meinekestrasse 4
D-1000 Berlin 15
Germany

Blackwell MZV
Feldgasse 13
A-1238 Wien
Austria

First published 1993

Set by Setrite Typesetters, Hong Kong
Printed and bound in Great Britain
at The Alden Press, Oxford

DISTRIBUTORS

Marston Book Services Ltd
PO Box 87
Oxford OX2 0DT
(*Orders*: Tel: 0865 791155
Fax: 0865 791927
Telex: 837515)

USA
Blackwell Scientific Publications, Inc.
238 Main Street
Cambridge, MA 02142
(*Orders*: Tel: 800 759–6102
617 876–7000)

Canada
Oxford University Press
70 Wynford Drive
Don Mills
Ontario M3C 1J9
(*Orders*: Tel: 416 441–2941)

Australia
Blackwell Scientific Publications Pty Ltd
54 University Street
Carlton, Victoria 3053
(*Orders*: Tel: 03 347–5552)

A catalogue record for this title
is available from the British Library

ISBN 0-632-03544-7

Library of Congress
Cataloging-in-Publication Data

Aeolian sediments/edited by
K. Pye and N. Lancaster.
p. cm.
(Special publication number 16
of the International Association of Sedimentologists)
Includes bibliographical references
and index.
ISBN 0-632-03544-7
1. Wind erosion. I. Pye, Kenneth.
II. Lancaster, N.
III. Series: Special publication number ...
of the International Association of Sedimentologists;
no. 16.
QE597.A35 1993
551.3′7—dc20

Contents

Preface

Studies of aeolian sediments, both ancient and modern, have exhibited a number of important conceptual advances in recent years. In particular, there has been a move away from descriptions of sediments, bedforms and sedimentary environments towards a new emphasis on the dynamics of aeolian depositional systems at different temporal and spatial scales and their response to external changes in sea levels, regional and global climates, and tectonics.

This Special Publication contains a selection of papers which were presented at the symposium entitled 'Aeolian Sediments, Ancient and Modern' held during the Thirteenth International Sedimentological Congress in Nottingham (August 1990). We have also added a number of papers from those who were not able to attend the meeting, but whose work represents important aspects of contemporary research in aeolian sedimentology.

A full understanding of aeolian sedimentary systems requires knowledge of the nature of the operative processes ranging from the scale of the individual grain to that of the regional sand sea. Continuing work on the mechanics of sand entrainment by wind is exemplified in the paper by Willetts & McEwan, while interactions between wind, sand transport and bed morphology at the dune scale are illustrated by Burkinshaw & Rust's field studies in the modern Alexandria coastal dunefield of South Africa. The controls on aeolian processes and their effects at the dunefield and regional scales are illustrated in the papers by Pye and Corbett, which present data from active systems on the Queensland coast and in the Namib Desert, respectively. Both of these papers demonstrate that sand accumulation in these areas has a long history, and evolution of the dunefields has been greatly influenced by changes in climate and sea level.

New techniques can contribute significantly to the recognition of genetic units in modern aeolian accumulations. Particularly exciting in this context is the recent application of ground-penetrating radar to studies of sedimentary structures at Great Sand Dunes, Colorado, by Schenk and his co-workers.

It is now becoming increasingly clear that the development of most thick aeolian sediment sequences, both relatively recent and ancient, has been episodic, in response to external changes that affect sediment supply and bedform mobility. The key to understanding the episodic nature of aeolian accumulation has been provided by the recognition that major bounding surfaces (supersurfaces) separate genetic units in the deposits of many ancient and modern sand seas. Supersurfaces truncate aeolian deposits and therefore represent periods when aeolian accumulation ceased, sometimes to be replaced by deflation or fluvial erosion over large areas of a sand sea. Recognition and tracing of super bounding surfaces formed by different processes provide the basis for constructing aeolian event stratigraphies and for the reconstruction of the depositional history of both modern and ancient sand seas, as the papers by Lancaster and Havholm and co-workers demonstrate, respectively. An understanding of the spatial changes in the geomorphological and sedimentological character of aeolian genetic units (e.g. variations in dune type, grain size trends and changes in the proportions of non-aeolian facies present) also provides an important tool for the unravelling of the evolution of sand seas. Langford & Chan provide criteria for evaluating the position of genetic units within a sand sea, using examples from the Permian Cedar Mesa Sandstone.

Determination of the causes of episodic behaviour in aeolian sedimentary sequences presents perhaps the greatest challenge in aeolian sedimentology, and requires careful evaluation of the effects of changes in climate and sea level on sand supply, water table levels, vegetation cover and wind regime. Atchley & Loope show how cyclic marine sedimentation affected aeolian sedimentation in the Carboniferous Paradox Basin of southeast Utah, while Chrintz and Clemmensen demonstrate how careful investigation and modelling of sedimentary structures can be used to reconstruct aeolian process environments and to document their changes over time, thereby allowing correlation with other geological evidence of environmental change.

There can be little doubt that studies of active and late Quaternary aeolian sedimentary systems provide important information which can be used to understand ancient aeolian deposits. Equally, careful studies of thick aeolian sediment sequences in the rock record, which allow a detailed three-dimensional picture of sand sea structure to be built up and long-term evolutionary trends to be identified, provide a context within which the recent and likely future development of modern sand seas can be framed. Without doubt, this cross-fertilization of ideas between 'modern' and 'ancient' in aeolian sedimentology will continue to great effect.

We would like to thank all those who assisted with the task of reviewing the manuscripts in accordance with the standards of *Sedimentology*. Special thanks are due to G. Blount, S.G. Fryberger, J.E. Marzolf, M.L. Porter, V.P. Tchakerian, K. Glennie, G. Butterfield, H. Tsoar and D.S.G. Thomas.

K. PYE
Reading, UK

N. LANCASTER
Reno, Nevada, USA

Modern Aeolian Environments

Spec. Publs Int. Ass. Sediment. (1993) **16**, 3–11

Particle dislodgement from a flat sand bed by wind: a re-analysis of Willetts & Rice's data

B.B. WILLETTS *and* I.K. McEWAN

Department of Engineering, University of Aberdeen, King's College, Aberdeen AB9 2UE, UK

ABSTRACT

A number of numerical simulations of aeolian saltation have been constructed in the last few years, each requiring independent checks to ensure its validity. The object of this paper is to demonstrate compatibility between data calculated by one of these saltation models (McEwan & Willetts, 1991) and experimentally derived data from tracer grain experiments. The experiments are discussed with reference to two numerical simulations. Firstly, the depletion of dyed grains from a coloured strip is modelled, enabling dislodgement rates to be calculated. This simulation is a reworking of dislodgement data previously published by Willetts & Rice (1988); an important component in this reworking is an exponential reptation length distribution derived from the saltation model of McEwan & Willetts (1991). Importantly, the revised values of dislodgement rate calculated show closer agreement with other published work (Anderson, 1986; Jensen & Sorensen, 1986; Williams *et al.*, 1990). Secondly, the progress of the coloured grains toward a creep trap placed downwind of the coloured strip is modelled, again using an excursion length distribution calculated by the saltation model. A reasonable match is found between the calculated and experimentally measured grain arrival curves.

INTRODUCTION

Recent advances in the understanding of aeolian transport have clarified the role of inter-grain collision and, in general terms, of the change induced by saltation in the near-bed wind (Anderson *et al.*, 1991). The conspicuous consequence of this change is that transported grains of compact shape at wind speeds above threshold are mobilized predominantly by saltating grains colliding with the bed (impact dislodgement), while direct dislodgement by the wind plays a minor role. Numerical models incorporating calculated trajectories, statistical treatments of collision and elementary fluid mechanics of the interaction between saltating grains and the wind have demonstrated the attainment of a transport rate in equilibrium with the driving wind because of the modification of the boundary layer induced by the load of transported grains (Werner, 1990; Anderson & Haff, 1991; McEwan & Willetts, 1991). The calculation has been made only for material of uniform grain size, because the ejecta produced by collision have not been identified by grain size experimentally, so the 'splash functions' used to portray the effects of collision in numerical models contain no information about grain size.

It can be argued that numerical modelling, having lent considerable confidence to physical interpretations during the last 5 years, is on the point of outstripping the observational base on which it depends. There is an urgent need for confirmation by observation that the equilibrium rates of dislodgement, the excursion length distribution of dislodged grains and the wind speed distribution arrived at by the models are realistic. Moreover, the library of data available is inadequate for the purpose of elucidating the influence on the transport process of the distribution of grain sizes present in the parent (bed) material. Since surface resorting is associated with rippling, and since both are universal features of aeolian activity, it is clearly important to quantify the role of size distribution and perhaps to develop a generation of models which incorporate a hypothesis about its role. The existing data have been acquired with great difficulty, and acquiring the necessarily more detailed observations which are now required will be even more difficult. Therefore it seems likely that evidence will be accumulated of those details which can be observed and have an implicit bearing on the matter, rather than of those which, were they attainable, would be most effective in resolving uncertainty.

The present study reworks grain dislodgement data first shown in Willetts & Rice (1988). Moreover, it introduces previously unpublished data (obtained from the same study) on the mean progress velocity of grains. A reconciliation is sought between the two data sets, using an estimate of path length distribution derived from a numerical model of the saltation cloud (McEwan & Willetts, 1991). The data came from wind-tunnel experiments using a quartz dune sand with a natural size distribution (Fig. 1); the saltation model employs uniform grains of the modal size of this distribution.

If r is the dislodgement rate per unit area and time, and p_λ the probability that a dislodged grain will have excursion length λ, then the transport rate, q_s, of grains in a uniform flow field can be expressed thus:

$$q_s = \int_{\lambda=0}^{\lambda_{max}} \int_{x=0}^{x=\lambda} p_\lambda r \, dx \, d\lambda \qquad (1)$$

where x is distance measured downwind.

Thus the transport rate can be determined if the dislodgement rate and the excursion length distribution are known. It is reasonable to suppose that both of these will be different for different size fractions in the population, so equation (1) should be evaluated fraction by fraction. If n fractions are distinguished,

$$q_s = \sum_{i=1}^{n} \int_{\lambda=0}^{\lambda=\lambda_{max}} \int_{x=0}^{x=\lambda} p_\lambda r \, dx \, d\lambda \qquad (2)$$

where i represents the fraction counter.

Not only the transport rate, but also the transported load and its distribution, depend on dislodgement rate and excursion length distribution. Anderson (1987) has demonstrated the influence of excursion length distribution on ripple growth. It is a straightforward inference from his argument that differences between such distributions for different size fractions will result in differential surface sorting along the ripple wavelength. Because of this pervasive influence, it is important that well justified values can be assigned to the two variables in discussion and simulation of the transport process.

There are a small number of sources of information about the two parameters. Jensen & Sorensen (1986) used estimated values of dislodgement rate, r, and launch velocity in a detailed simulation of the wind-tunnel experiments of Williams (1964). The estimates were modified by trial until reasonable agreement was achieved with Williams's observations of flux profile. Anderson (1986) also explored dislodgement rate by estimating the value required to render saltation layer calculations com-

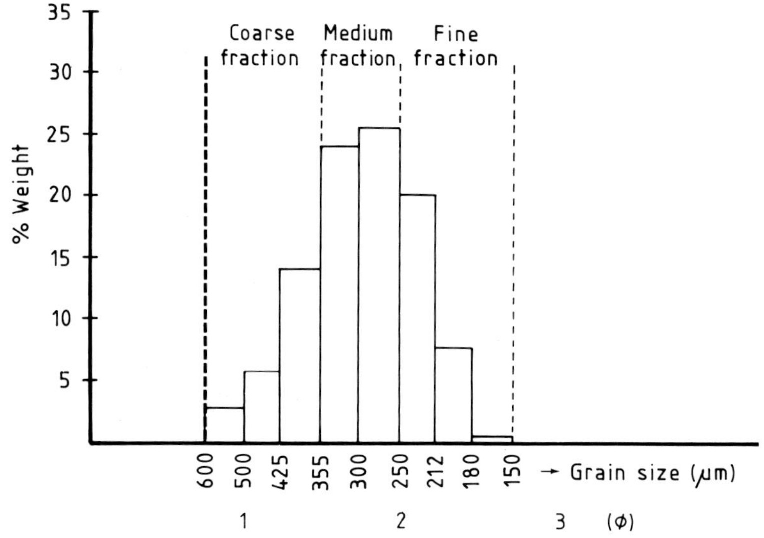

Fig. 1. Size distribution of the sand and identification of the coarse, medium and fine fractions.

patible with observed transport rates. Willetts & Rice (1988) made direct observations of the depletion rate of dyed grains from a pre-laid strip and estimated the dislodgement rate from the observed depletion rate. The estimation procedure involved an assumption about the distribution of excursion length; the distribution was assumed to be linear, very short excursions being the mode. For values of shear velocity, U_*, between threshold and 50 cm s^{-1}, the Jensen & Sorensen and the Willetts & Rice estimates are of the same order of magnitude: the Anderson values of dislodgement rate are an order smaller. Williams *et al.* (1990) investigated rates of aerodynamic entrainment from narrow transverse strips of loose grains placed at three positions up to 130 cm from the leading edge of a flat plate which had been roughened by the attachment of identical grains. They found that aerodynamic dislodgement rate decreased with distance from the leading edge of the flat plate and concluded that this was due to an increasing skewness in the instantaneous shear stress distribution, resulting in a lower frequency of events in which the instantaneous shear stress was greater than the threshold shear stress for the sediment. Moreover Williams *et al.* found *aerodynamic* dislodgement rates an order of magnitude higher than the *impact* dislodgement rates found by Willetts & Rice (1988). They suggested that this apparently paradoxical result was caused by two factors: firstly, that replenishment in the experiments of Willetts & Rice (1988) caused the impact dislodgement rate to be underestimated; and secondly, that the heterogeneous size distribution in the bed used by Willetts & Rice suppressed dislodgement. This paper will rework the original data of Willetts & Rice (1988) using a revised excursion length distribution calculated by the saltation model of McEwan & Willetts (1991) to recalculate the influence of replenishment.

The present generation of saltation models provide further estimates of dislodgement rate. Anderson & Haff (1991) calculate a dislodgement rate of 1000 grains cm^{-2} s^{-1} for 250 μm grains at $U_* = 50$ cm s^{-1}. Sorensen (1991) calculates the dislodgement rate to be 1540 grains cm^{-2} s^{-1} for 300 μm grains at the same wind speed while the saltation model of McEwan & Willetts (1991) calculates the dislodgement rate to be 560 grains cm^{-2} s^{-1} at $U_* = 50$ cm s^{-1} for 300 μm grains. These calculated values of impact dislodgement rate are broadly consistent and are approximately an order of magnitude larger than those estimated by Willetts & Rice (1988).

This paper will rework the data of Willetts & Rice (1988) using a distribution of excursion length revised by reference to a numerical model of the transport process. The resulting dislodgement rate together with the adopted excursion length distribution will then be used to calculate grain progress, which will be compared with observations of grain progress made by means of a trap placed downwind of the source of coloured grains. These observations were made during the same experiments as those described in Willetts & Rice (1988).

EXPERIMENTAL PROCEDURE AND RESULTS

Experiments were performed in a wind-tunnel of cross-section 500 × 500 mm and length 10 m. Air was blown along the tunnel and, near the upwind end, was seeded with sand released into it at a few centimetres above the bed. Also at the upwind end of the tunnel a variable porosity grid was used to prematurely thicken the boundary layer on the tunnel floor. The arrangement is reported in more detail in an earlier paper (Phillips & Willetts, 1979).

The experiments discussed here are those in Willetts & Rice (1988). In these experiments the whole of the wind-tunnel floor was covered to a depth of 10 mm with a predominantly quartz sand obtained from local sand dunes. The size distribution, obtained by sieving, is shown in Fig. 1, which also indicates three fractions, nominated 'coarse', 'medium' and 'fine'. Sand taken from this population was dyed, as a means of identification in observations of grain movement. A thin cold water dye was used which, it was considered, would not measurably alter the grain properties. The sand was separated into the three indicated fractions, each of which was then dyed a different colour. The coloured fractions were then remixed in the proportions found in the parent sand (22.2% coarse, 49.8% medium and 28.0% fine by weight). This coloured sand was then spread over the 10 cm of uncoloured sand to a thickness of approximately one grain in a straight transverse strip situated 6 m from the upwind end of the tunnel, spanning the tunnel width, and 100 mm in extent in the flow direction (Fig. 2). The intention was that, in this strip, each small bed area should contain coloured grains, and no uncoloured ones should be visible in a plan view of the strip.

A camera mounted above the tunnel roof was focused on an area of the strip near its upwind edge

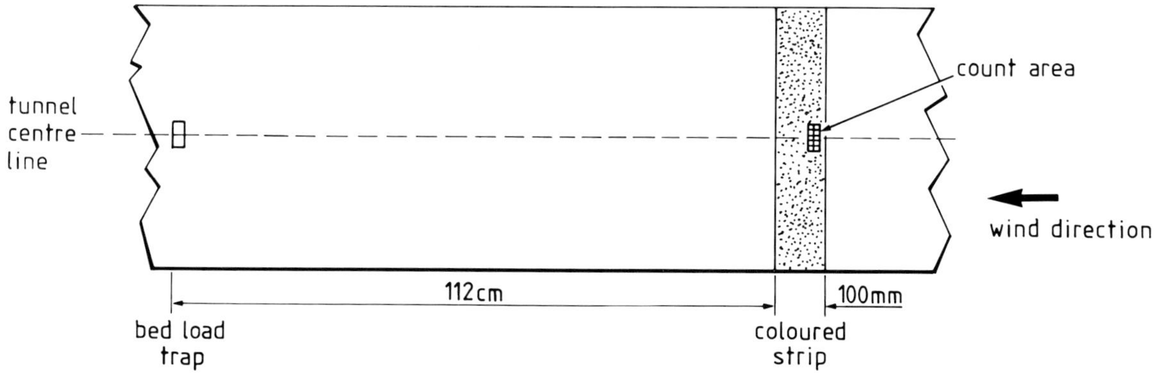

Fig. 2. The designated area for counting relative to the placed strip of coloured grains and the position of the bed-load trap shown downwind of the strip.

and on the centre line of the tunnel floor. It was used to take coloured slide photographs before the start of each experiment and at 5 s intervals during the experiment. For the first of these exposures (with still air in the tunnel) a rectilinear grid of 10 mm squares outlined in cotton was placed on the sand in the target area. The image of this grid was used as an aid to grain counting in this photograph and the others of the same series, each series representing one experiment in which camera position and setting were unaltered.

A bed-load trap (Fig. 3) was installed in the floor of the wind-tunnel on its centre line and 112 cm from the downwind edge of the coloured strip. Grains progressing as bed-load within its effective width were trapped; while the source of the undyed grains was unspecific, the origin of trapped coloured grains was evident. The trap was designed and operated so that the time of arrival of each coloured grain could be estimated (the front of the trap was transparent, and time horizons were drawn on it at 5 s intervals to record the progress of sand accumulation).

Four experiments were performed, at shear velocities of 24, 32, 39 and 48 $cm\,s^{-1}$. In each of them the removal rate of coloured grains from the pre-laid strip and the pattern of arrival of coloured grains at the trap were both recorded. Surviving grains on the coloured strip were counted on the photograph taken at each time interval, with the assistance of the photographed grid mentioned above. The arrival time of each grain at the trap was estimated by the interpolation between the lines drawn at known time intervals to record the accumulation surface.

The results for the coarse fraction are summarized in Figs 4 and 5. (Depletion rates for medium and fine fractions for the same experiments are given in Willetts & Rice (1988).)

Figure 5 shows that the largest number of grains enter the trap at $U_* = 39\,cm\,s^{-1}$. Moreover, by coincidence, there were fewer grains present on the tracer strip at $U_* = 39\,cm\,s^{-1}$ than at the other shear velocities (see table 1, Willetts & Rice, 1988). The present authors advance the following reasoning for this result. The depletion curves for the coloured strip show that at low shear velocities some coloured grains remained on the strip after 90 s. This suggests a relatively long residence time on the strip and, by inference, a slow progression towards the bed-load trap located downwind. Thus the small catch of grains made at $U_* = 24\,cm\,s^{-1}$ (see Fig. 5) was expected. As the shear velocity is increased the residence time (on the strip) is decreased and more grains reach the trap during the period of the experiment. However, it is plausible that at the highest shear velocity many more coarse grains are entering successive saltation, participating in longer hops, and thus pass over the trap. Note that although there are fewer grains caught in the bed-load trap at $U_* = 48\,cm\,s^{-1}$ than at $U_* = 39\,cm\,s^{-1}$, those grains that are caught have made a more rapid progression towards the trap. This argument may suggest a transition or threshold between $U_* = 39\,cm\,s^{-1}$ and $U_* = 48\,cm\,s^{-1}$, above which the majority of coarse grains begin participating in saltation and below which the coarse grains make relatively slow progress along the surface in shorter loops.

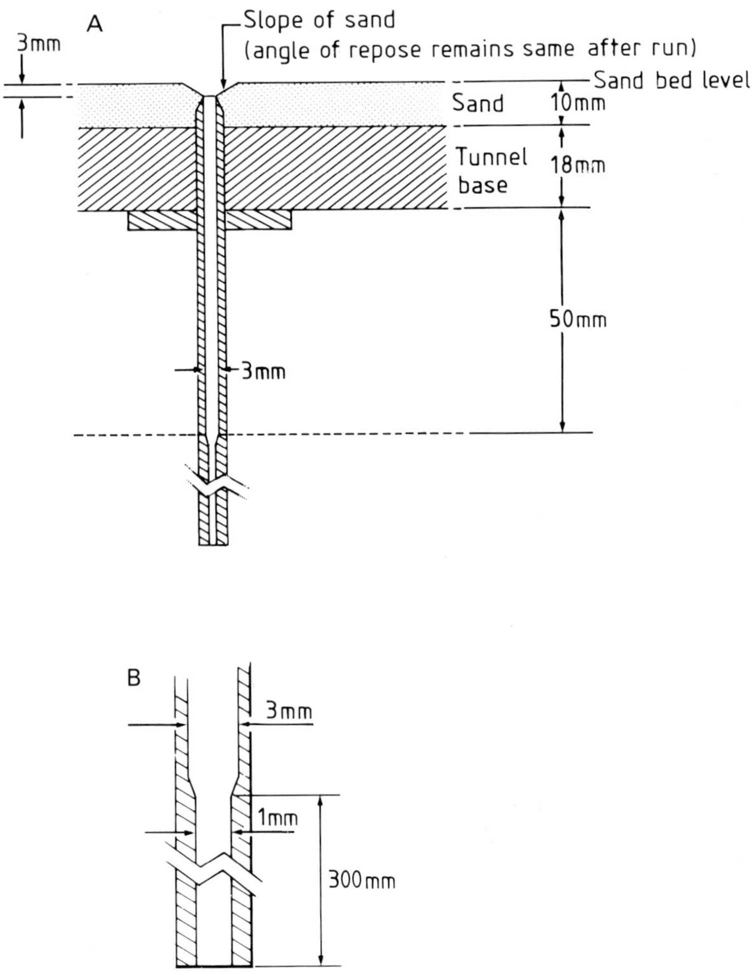

Fig. 3. (A) Sketch of the bed-load trap without the collector box. (B) A section showing the trap in relation to bed level.

NUMERICAL SIMULATIONS

Dislodgement rate

It can readily be shown (Willetts & Rice, 1988) that the net rate of removal of coloured grains from a small area at a distance x from the upwind edge of a strip of such grains is given by

$$\frac{-\mathrm{d}N_{x,t}}{\mathrm{d}t} = kN_{x,t} - \int_{\lambda=0}^{\lambda_{max} \text{ or } x} p_\lambda \frac{\mathrm{d}N}{\mathrm{d}t}(x-\lambda)t \,\mathrm{d}\lambda \quad (3)$$

Here $N_{x,t}$ is the surviving number of grains at position x and time t, k is the dislodgement rate time constant, λ is excursion length, having maximum value λ_{max}, and p_λ is defined as above. The last term in equation (3) represents replenishment at position x of coloured grains from the population between the upwind edge of the coloured strip and x. The effect of this term, as its form implies, is influenced by the excursion length distribution.

There is no secure evidence as to the form of this distribution. Willetts & Rice (1988) considered following Anderson (1987) and using a gamma distribution to represent the distribution of excursion lengths. However, they believed that very short excursions were much more common than the gamma distribution would indicate and used instead a triangular distribution in which very short excursions were the mode.

There remains considerable doubt about both the grain dislodgement rate, and the excursion length distribution for a given grain population and wind condition. However, numerical models

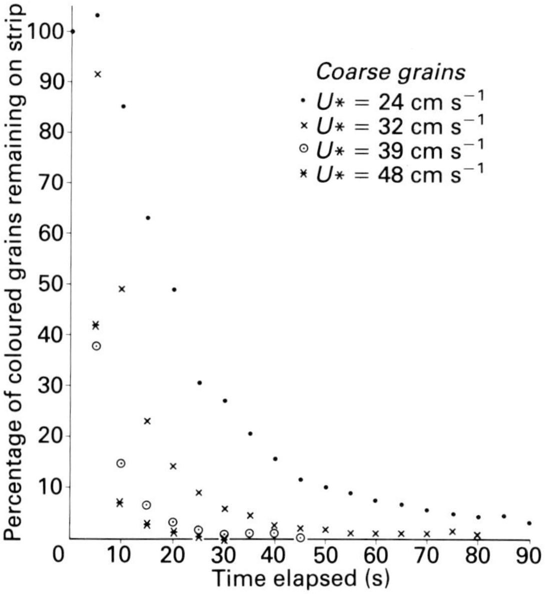

Fig. 4. The depletion of the coloured strip for the coarse fraction at four U_* values, from Willetts & Rice (1988).

of the equilibrium saltation layer which have now become available (Werner, 1990; Anderson & Haff, 1991; McEwan & Willetts, 1991) provide values, albeit surrogate ones, for these variables. A distribution of excursion length for dislodged grains generated by the model of McEwan & Willetts (1991) is shown in Fig. 6. The number of very short excursions is so much larger than was assumed in earlier investigations that doubt is cast on the significance of the replenishment terms in equation (3) in the determinations of dislodgement made in Willetts & Rice (1988). We have recalculated dis-

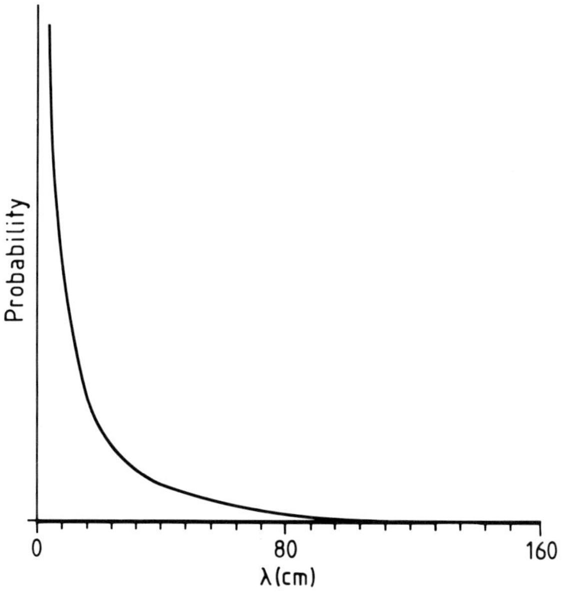

Fig. 6. Excursion length distribution for ejected grains generated by the saltation model at $U_* = 30\,\mathrm{cm\,s^{-1}}$.

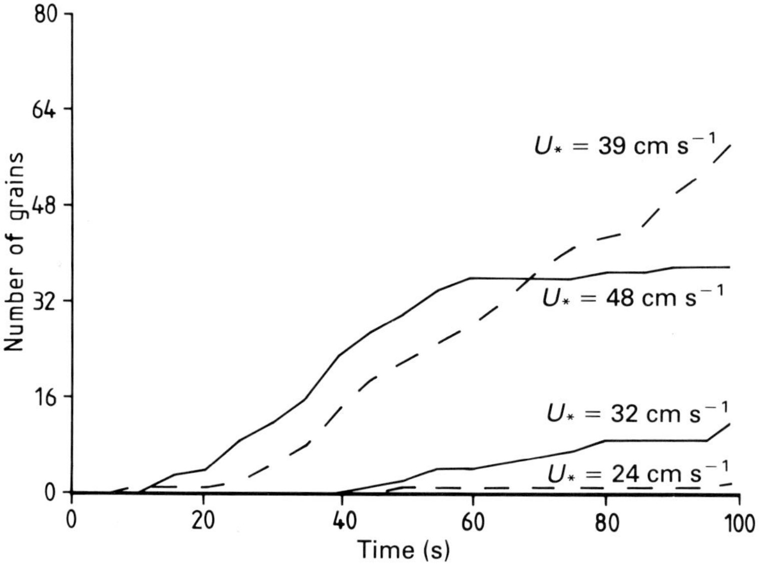

Fig. 5. The arrival of the coarse fractions at the bed-load trap at four U_* values. Note that the cumulative number of collected grains is shown.

lodgement rates using a discretized form of equation (3), and employing this modelled excursion length distribution, arriving at values which are plotted in Fig. 7 together with dislodgement rates calculated by the saltation model of McEwan & Willetts (1991). The exponent in the relationship between dislodgement rate and shear velocity is 2.6. This is subject to a good deal of uncertainty because of the limited data: the corresponding exponent for the dislodgement rate internal to the saltation model is also 2.6. The values of dislodgement rate are much higher than those estimated by Willetts & Rice (1988) because the excursion length distribution assumed in that study caused replenishment to be underestimated, as has been suggested by Williams *et al.* (1990) and Sorensen (1991). However, the experimental values of dislodgement are still lower than those calculated by the saltation model. The calculated values are for a 300 μm uniform sand whereas the experimental values are for a well graded natural sand (Fig. 1). This may well support the observation of Williams *et al.* (1990) that dislodgement rates are likely to be less in beds of heterogeneous grain size because of increased sheltering, shape variation, threshold ranges and more complex support geometry. Thus the need to develop models of saltation incorporating multiple grain size is emphasized.

At this point, we have an excursion length distribution and a revised estimate of dislodgement rate which, together, make it possible to track, probabilistically, the progress of grains along the wind-tunnel. Thus, a simulation of the observations of coloured grains collected in the trap can be attempted.

Grain progress towards the trap

The simulation of grain progress from the coloured strip to the trap is reported for the coarse fraction only. The calculation advances in 1 s steps, remapping grain position at each step. At each step, each grain is assigned a random number which determines whether it is to move or not; if it does move, then it is allocated an excursion length in accordance with the calculated distribution of excursion lengths shown in Fig. 8. This excursion length distribution is based on the full distribution of saltation path lengths generated by a saltation model (McEwan & Willetts, 1991) and hence it incorporates the distribution shown in Fig. 6. Grains making an excursion of 27 cm or more were assumed to enter a sequence of repeated saltations and to be lost to the calculation. The figure of 27 cm is arbitrary, but the assumption accords with the general observations of Tsuchiya &

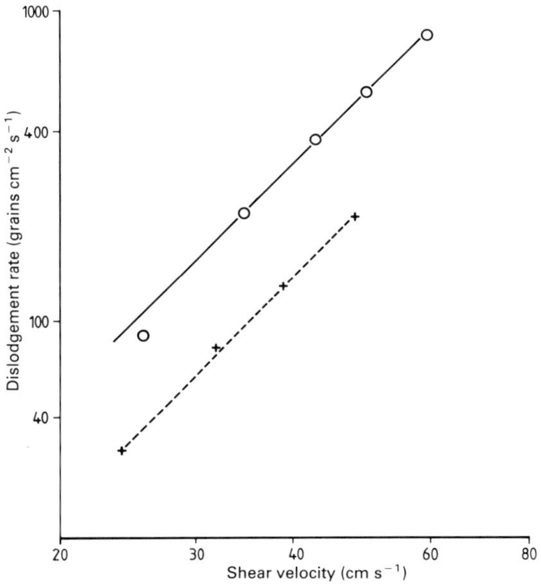

Fig. 7. Dislodgement rate shown against shear velocity: (+) derived from tracer experiment; (○) calculated from saltation model of McEwan & Willetts (1991).

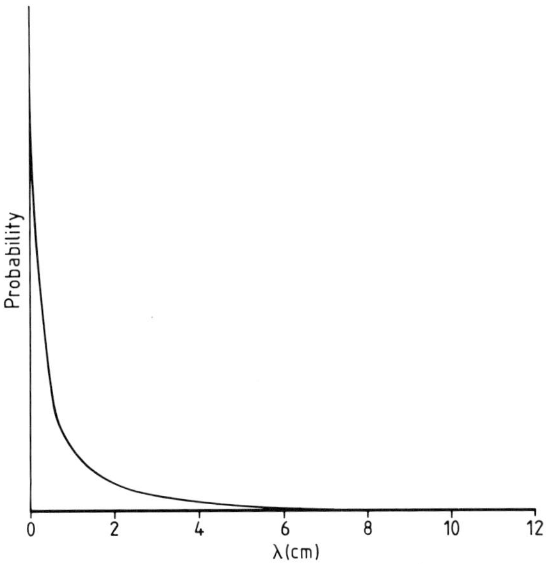

Fig. 8. The excursion length distribution for the full grain population at $U_* = 48\,\mathrm{cm\,s^{-1}}$.

Kawata (1972) and Mitha *et al.* (1986) and the impact simulations of Anderson & Haff (1991).

Results of the simulation are shown in Fig. 9 for the experiment conducted at $U_* = 48\,\mathrm{cm\,s^{-1}}$.

SUMMARY AND CONCLUSIONS

The recalculation of dislodgement rates from the data of Willetts & Rice (1988) using a revised estimate of the excursion length distribution calculated by the saltation model of McEwan & Willetts (1991) results in improved agreement with previously published dislodgement data (Anderson, 1986; Jensen & Sorensen, 1986; Williams *et al.*, 1990). Moreover, there is some agreement between the experimental results and the saltation model of McEwan & Willetts (1991) and the form of the relationship between dislodgement rate and shear velocity. However, differences remain between magnitudes of dislodgement rate found experimentally by Williams *et al.* (1990) and by the saltation models of Anderson & Haff (1991) and McEwan & Willetts (1991) and those calculated from the experimental results of Willetts & Rice (1988). The lower dislodgement rates found in the latter study may in part result from the heterogeneous grain size bed used in their experiments: the other results are for a uniform size sand. Thus the need to develop saltation models incorporating multiple grain size is emphasized.

Now that transport rates are being calculated by means of models which simulate the physics of the transport process, it is necessary to establish control by checks on the rate at which subprocesses proceed. There is a dearth of competent data for such checks. This study has sought to check the mutual consistency of a calculated distribution of excursion lengths with observed shifts of location of a population of colour-tagged grains.

The overall success of the simulation of grain progress is impressive (Fig. 9). However, there is more than one combination of the assumed excursion length distribution and interpretation of the loss rate from the coloured strip (Figs 4 and 5) which would give good results. Therefore internal checks are important to supplement the check on the finished prediction of progress to the trap.

One such check is that the delay before grains start arriving at the trap is forecast reasonably correctly by the calculation based on the distribution of excursion lengths (Fig. 8). This, together with a correctly timed build-up to the full catch of coloured grains, and the dependency of the calculated dislodgement rate on the excursion length distribution, provide quite strong evidence for the validity of the simulation.

There is, however, an instructive qualification of the degree of success achieved. Figure 10 illustrates

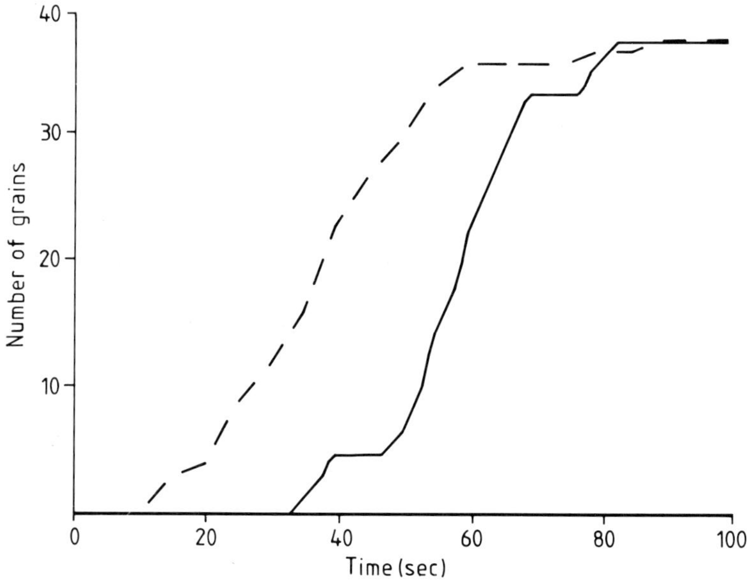

Fig. 9. The arrival of coarse grains at the bed-load trap at $U_* = 48\,\mathrm{cm\,s^{-1}}$. The solid line indicates the simulation results and the dashed line denotes the experimentally measured curve.

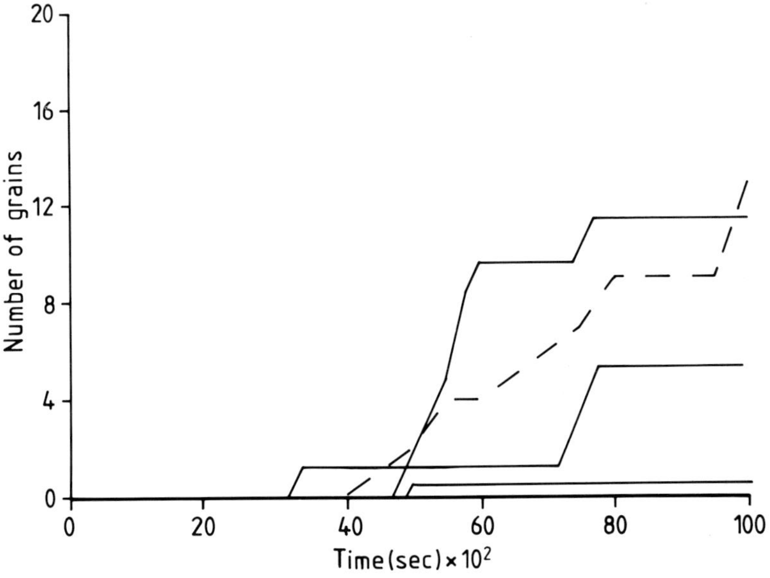

Fig. 10. The arrival of coarse grains at the bed-load trap at $U_* = 32\,\text{cm}\,\text{s}^{-1}$. The dashed line indicates the experimentally measured curve and the solid curves represent three separate simulations, the only difference being in the sequence of random numbers chosen in the calculation (see Summary and conclusions).

the outcome of repeated simulations of the same experiment. The procedure is identical; all that is changed is that the random number sequence from which the dislodgement is determined is entered at a different point. The 'numerical experiment' has a different outcome for different random number sequences. Since we believe that the number of coloured grain movements in the simulations conforms reasonably to nature, it seems evident that a number of physical experiments at the same wind speed would also show different results.

Considerable caution is still required, therefore, in interpreting these results. However, we believe that the evidence is strong enough to justify further tracer experiments of the same general type, undertaken with the specific objective of providing data with which to check the soundness of numerical simulations of grain transport.

REFERENCES

Anderson, R.S. (1986) Erosion profiles due to particles entrained by wind: application of an eolian sediment-transport model. *Geol. Soc. Am. Bull.* **75**, 1270–1278.

Anderson, R.S. (1987) A theoretical model for aeolian impact ripples. *Sedimentology* **34**, 943–956.

Anderson, R.S. & Haff, P.K. (1991) Wind modification and bed response during saltation of sand in air. *Acta Mechanica Suppl.* **1**, 21–52.

Anderson, R.S., Sorensen, M. & Willetts, B.B. (1991) A review of recent progress in our understanding of aeolian sediment transport. *Acta Mechanica Suppl.* **1**, 1–20.

Jensen, J.L. & Sorensen, M. (1986) Estimation of some aeolian saltation transport parameters: a re-analysis of Williams' data. *Sedimentology* **33**, 547–558.

McEwan, I.K. & Willetts, B.B. (1991) Numerical model of the saltation cloud. *Acta Mechanica Suppl.* **1**, 53–66.

Mitha, S., Tran, M.Q., Werner, B.T. & Haff, P.K. (1986) The grain bed impact process in aeolian saltation. *Acta Mechanica* **63**, 263–278.

Phillips, C.J. & Willetts, B.B. (1979) Predicting sand deposition at porous fences. *J. Waterways Port, Coast, Ocean Div., Am. Soc. Civ. Engs* **105**(WW1), 15–31.

Sorensen, M. (1991) An analytic model of wind-blown sand transport. *Acta Mechanica Suppl.* **1**, 67–82.

Tsuchiya, Y. & Kawata, Y. (1972) Characteristics of saltation of grains by wind. *Proc. 13th Int. Coastal Eng. Conf.*, 1617–1625.

Werner, B.T. (1990) A steady-state model of wind blown sand transport. *J. Geol.* **98**, 1–17.

Willetts, B.B. & Rice, M.A. (1988) Particle dislodgement from a flat sand bed by wind. *Earth Surf. Proc. Landf.* **13**, 717–728.

Williams, G. (1964) Some aspects of the eolian saltation loads. *Sedimentology* **3**, 257–287.

Williams, J.J., Butterfield, G.R. & Clark, D.G. (1990) Rates of aerodynamic entrainment in a developing boundary layer. *Sedimentology* **37**, 1039–1048.

Spec. Publs Int. Ass. Sediment. (1993) **16**, 13–21

Aeolian dynamics on the windward slope of a reversing transverse dune, Alexandria coastal dunefield, South Africa

J.R. BURKINSHAW *and* I.C. RUST

Institute for Coastal Research, Department of Geology, University of Port Elizabeth, PO Box 1600, Port Elizabeth 6000, South Africa

ABSTRACT

Few empirical measurements describing the inter-relationship between wind shear, sediment flux and dune shape have been reported. We compare the results from two of several detailed field experiments conducted as part of a study of airflow on the windward slope of a 7 m high reversing transverse coastal dune. The seasonal reversal of the dune in response to opposing summer and winter wind regimes provides an ideal opportunity to study the relationship between dune shape and airflow characteristics.

The wind speed profile, erosion and deposition rates and values of surface gradient were recorded along a section line normal to the crest line of the dune. Our paper deals specifically with wind speed measured at 6 cm above the surface. The two experiments were conducted 3 weeks apart at the start of the summer wind season on the east flank of the dune. Wind conditions were similar but the difference in the crestal geometry of the dune resulted in contrasting erosional responses.

Variation in sediment flux can be related not only to the absolute gradient of the dune surface but also to changes in gradient. We recognize three phases in the erosional response of the dune which are determined by change in the rate of change of gradient, and which give an insight into the self-regulatory relationship between form and flow. Erosion pins act as a sensitive measure of variation in shear stress on the dune surface, and could possibly be used in future shear stress determination once a data base of variables related to shear stress has been acquired.

INTRODUCTION

The intrusion of an aeolian bedform into the atmospheric boundary layer results in the acceleration of flow on the upstream side of the dune form. The extent of the influence of the form on the flow and the self-regulatory response of aeolian bedforms to the compression of airflow streamlines have yet to be evaluated.

Variation in sediment flux on the windward slope occurs because of changes in surface shear stress, and changes in the gradient and orientation of the slope relative to the oncoming wind (Howard *et al.*, 1978; Lai & Wu, 1978; Watson, 1987). Few empirical measurements describing the inter-relationship between wind shear stress, sediment flux and dune shape have been reported. Howard *et al.* (1978) successfully related sand transport to wind measurements on a barchan dune in a study supplemented by wind-tunnel observations. Several studies have made an attempt to relate sand transport to wind speed and/or direction (Tsoar, 1983; Lancaster, 1985; Livingstone, 1986).

The most detailed study of transverse dune dynamics published so far is an experiment examining aerodynamics and sand transport on a 1 m high transverse coastal dune (Hesp *et al.*, 1989); variation in sand transport rates was related to variation in wind velocity, but the shape of the dune was not taken into account.

Bagnold (1941) related the rate of sand removal, dQ/dx, to the tangent of the angle of inclination of the dune slope, ϕ:

$$\frac{dQ}{dx} \propto \gamma \tan\phi$$

where γ is the specific weight of sand in bulk. This is a geometrical relationship based on dune shape, and indicates that there is neither erosion nor deposition at the dune crest, and that the rate of removal is greatest where the local slope is steepest. As Bagnold himself recognized (p. 201), the equation is appropriate for preserving the curvature of the stoss slope, but neglects the variation of the wind force along the dune surface.

Sand transport rates are determined by surface shear stress, as measured by shear velocity, v^*, which for wind blowing over a mobile sand surface is defined as:

$$v^* = \frac{(v - v_t)}{5.75 \log(z/k')}$$

where z is the height at which v is measured. Bagnold (1941) found that for a naturally graded sand, v_t (threshold velocity) is $4\,\mathrm{m\,s^{-1}}$, and suggested that a value of 1 cm be assigned to k', which can be related to the height of wind ripples.

Sand transport equations have been satisfactorily developed for wind blowing over a flat surface. However, the development of a non-logarithmic wind speed profile above the windward slope of a dune makes it difficult to know what is an accurate measure of shear velocity on a dune surface and casts doubt on the estimation of sand transport based on Bagnold's equation (Mulligan, 1988). In addition, the rate of sand transport on a sloping surface is affected by the gradient of the surface, but as yet no applicable experimental data have been collected, and transport equations are corrected by available theory (Howard *et al.*, 1978; Lai & Wu, 1978).

Watson (1987, p. 516) advocated that 'since variations in shear stress are related to changes in flow velocity and pressure gradient, future work should attempt to elucidate the interaction of these three parameters'.

Our paper compares the results from two detailed field experiments conducted as part of a study of airflow on the windward slope of a 7 m high reversing transverse dune in the Alexandria coastal dunefield, South Africa (Fig. 1). The study dune was selected from a series of nearly identical small dunes on the western edge of the dunefield (Fig. 2). The regularity of both the shape and the spacing of these dunes suggests the existence of a well structured airflow over the dunes, which made the area seem a logical place to conduct measurements. The seasonal reversal of the dune morphology with respect to the summer easterlies and winter northwesterlies provides an ideal opportunity to study the inter-relationship between dune shape and airflow characteristics.

Wind velocity and rates of erosion, together with local gradient measurements, were recorded on the east flank of the dune on two separate occasions, 3 weeks apart, early in the easterly (summer) wind season, 1988. The east flank was in the process of being reversed from an avalanche slip face (developed during the preceding winter wind season) to a stoss slope typical of the summer wind season. The wind speed was of the order of $13\,\mathrm{m\,s^{-1}}$ at 140 cm height above the crest on both occasions; however, the difference in the crestal geometry of the dune for these 2 days resulted in contrasting erosional responses. We recognize three phases in the erosional

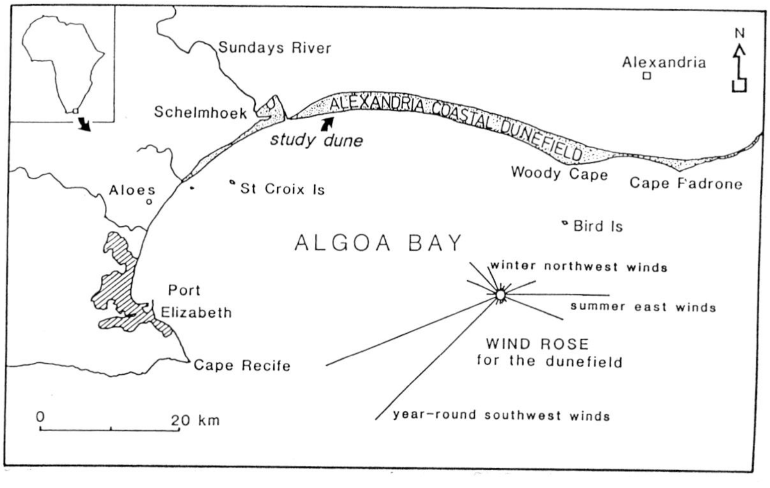

Fig. 1. Locality map of the Alexandria coastal dunefield including a stylized wind rose for the dunefield. (Modified from Illenberger & Rust, 1988.)

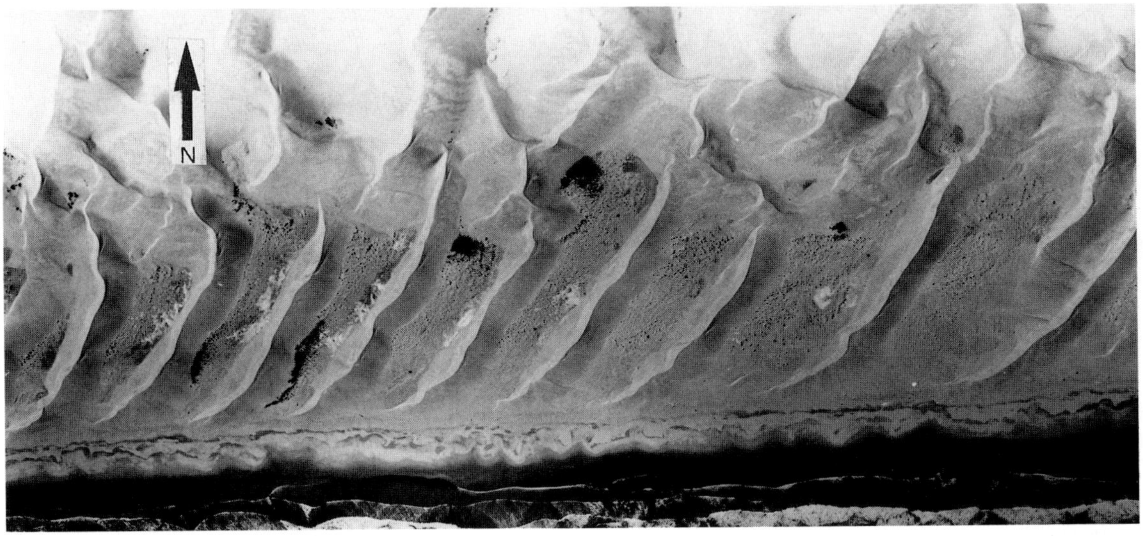

Fig. 2. Vertical aerial photograph of a 1 km stretch of the small reversing transverse dunes at the western edge of the dunefield, January 1989. The dunes are 150–300 m in length and dune spacing varies from 80 to 130 m. (Photograph: Werner K. Illenberger.)

response which correspond to changes in the rate of change of gradient of the windward slope.

METHODS

We measured airflow profiles on the windward slope along a number of transverse section lines normal to the dune crest; in this paper we refer exclusively to the data for one specific section line, known as SL60. Two anemometer arrays, each consisting of four micro-anemometers mounted at average heights of 6 cm, 20 cm, 60 cm and 140 cm above the dune surface, were deployed; one array was stationed permanently at the crest, while the other was deployed at different locations along SL60 measuring airflow for an average of 20 min per location.

A code was devised to facilitate reference to each anemometer relative to its position in three-dimensional space on the dune. The code consists of three parts, A/B/SLC, where A refers to the height in centimetres of the cups above the dune surface; B refers to the position of the array on the section line as a slope distance in metres from the base of the dune; and C identifies the section line. In cases where the entire array is referred to, the code is B/SLC.

In this paper, we provide data for the full measured wind speed profile but our discussion focuses on the data recorded at 6 cm above the surface. We plan to deal with the entire measured wind speed profile in another paper.

For a variety of technical reasons, the anemometers were not rigged at consistent heights above the dune surface for all measurement sessions, and for comparative purposes, where necessary, values of airflow speed are interpolated to the above-mentioned heights. Anemometers were initially cross-calibrated in the field and at the end of the project they were calibrated in a wind-tunnel.

Rates of erosion and deposition were monitored by means of erosion pins staked at 2 m intervals (on the second occasion we used 1 m intervals in the crestal zone) on the section line over a slope distance of 24 m, and the gradient was measured at the same locations using a Brunton clinometer. The location of an erosion pin in two-dimensional space on the dune surface is referred to using the same code devised for identifying the position of array stations.

The geometry of the windward slope was reconstructed by converting the gradient measurements and slope distance between measurements to height values at the corresponding horizontal distance from the base of the dune.

RESULTS

Wind speed data

For comparative purposes wind speed data at the various measuring stations are normalized with respect to the wind speed measured at 140 cm height above the dune crest (the top anemometer of the array stationed at the crest).

17 November 1988

The windward slope geometry of SL60 on 17 November 1988 (the beginning of the monitoring session) shows the rounded nature of the crest on that date (Fig. 3). Crestal wind speed as measured by the reference anemometer at 22/SL60 averaged a speed of 13.1 m s^{-1} between 12:00 and 14:00, and then dropped off gradually to a value of 10.0 m s^{-1} by 16:00.

The set of wind speed data (Table 1), based on all 20 anemometer stations, indicates that the wind speed measured (or interpolated) at 6 cm above the dune surface was fastest at mid-slope (anemometer 6/12/SL60). The airstream accelerated from 2.0 m s^{-1} to 8.1 m s^{-1} as it moved from the base towards mid-slope; this represents an increase of 40% of the wind speed normalized relative to the crestal wind speed. Airflow decelerated by 13% between mid-slope and the crest (air speed decreased from 8.1 m s^{-1} to 6.1 m s^{-1}).

Assuming a constant threshold value of 4 m s^{-1} at 1 cm height, calculated values of shear velocity within 6 cm of the surface show a corresponding increase on the lower slope from 0.37 m s^{-1} at 6/6/SL60 to 0.91 m s^{-1} at 6/12/SL60, and a corresponding decrease to 0.47 m s^{-1} at the crest (Table 1).

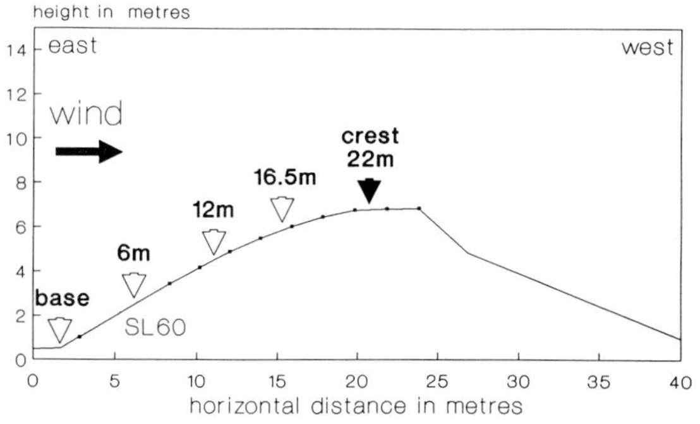

Fig. 3. Windward slope geometry of SL60, 17 November 1988. Open arrows indicate the station locations of the movable anemometer array.

7 December 1988

Reconstruction of the windward slope geometry 3 weeks later on 7 December 1988 shows that the crest had attained a triangular profile as a result of a relatively straight windward slope and a sharply peaked crest (Fig. 4). On this day the crestal wind speed averaged 13.0–13.7 m s^{-1} at 147 cm height over a 2 h time period (Table 2).

Table 1. Wind speed data as measured on 17 November 1988: figures in parentheses are shear velocity (m s^{-1}); other figures are velocity (m s^{-1})

	Anemometer height (cm)				
Station	6	20	60	140	v_{ref}[a]
22/SL60 11:58–14:04	6.1[b] (0.47)	10.5[b] (1.47)	12.8 (0.83)	13.1 (0.12)	13.1
16.5/SL60 12:12–12:47	7.0 (0.67)	12.1 (1.71)	12.3 (0.07)	14.1 (0.71)	12.7
12/SL60 13:27–14:04	8.1[b] (0.91)	11.2 (1.07)	11.9 (0.25)	12.2 (0.12)	13.5
6/SL60 14:56–15:37	5.7[b] (0.37)	7 (0.9)	7.6 (0.21)	7.9 (0.12)	11.1
0/SL60 16:07–16:21	2.0[b]	2.6 (0.2)	3.7 (0.48)	4.6 (0.36)	10.0

[a] Reference velocity measured at 140 cm above the crest.
[b] Interpolated value.

Table 2. Wind speed data as measured on 7 December 1988: figures in parentheses are shear velocity (m s^{-1}); other figures are velocity (m s^{-1})

	Anemometer height (cm)				
Station	6	20	60	140	v_{ref}[a]
21/SL60 14:36–14:54	9.1[b] (1.13)	12[b] (0.97)	13.4[b] (0.51)	13.7	13.7
17/SL60 14:04–14:24	6.5 (0.56)	10.7 (1.4)	12.1[b] (0.51)		13.1
13/SL60 14:36–14:54	6.6[b] (0.58)	10.1 (1.43)	11.4[b] (0.47)		13.7
8/SL60 15:17–15:45	6.5[b] (0.56)	8.8 (0.77)	9.7 (0.33)		13.7
0/SL60 16:04–16:15	2.6	3.4[b] (0.26)	4.7[b] (0.47)		13.0

[a] Reference velocity measured at 140 cm above the crest.
[b] Interpolated value.

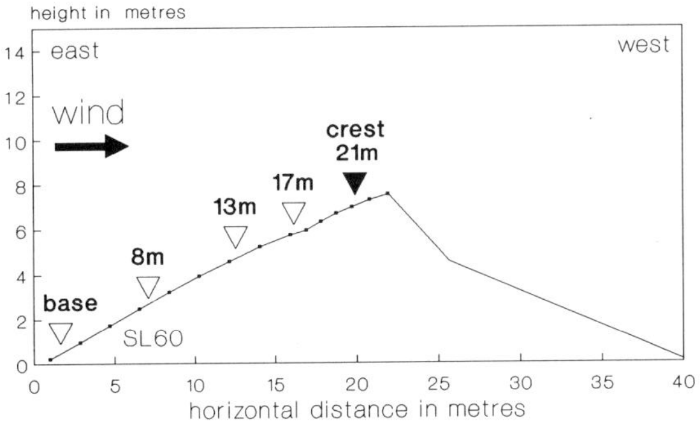

Fig. 4. Windward slope geometry of SL60, 7 December 1988. Open arrows indicate the station locations of the movable anemometer array.

The windflow pattern differed significantly from the flow recorded on 17 November. The wind speed at a uniform height of 6 cm above the dune surface showed a maximum value of 9.1 m s^{-1} at the crest (21/SL60); however, the rate of acceleration from the base of the slope towards the crest was not uniform. After the initial acceleration of the air speed on the lower slope of the dune from a value of 2.6 m s^{-1} at 0/SL60 (base of the dune) to a value of 6.5 m s^{-1} at 8/SL60, air speed remained more or less constant in the middle to upper regions of the slope until 17/SL60. Rapid acceleration occurred between this station and the crest.

Shear velocity calculated within 6 cm of the surface on the windward slope shows a sympathetic increase and decrease with the wind speed behaviour (Table 2). A maximum shear velocity of 1.13 m s^{-1} was calculated within 6 cm of the surface in the crestal area.

Erosion pin and gradient data

For comparative purposes, the rate of sediment flux is normalized in the text in terms of millimetres of vertical change in the dune surface per minute, but actual vertical change measured at each erosion pin over a given period of measurement is plotted in the diagrams (Figs 5 and 6).

17 November 1988

The erosion pin data set is subdivided into two time periods corresponding to the initial period of strong wind (11:30–14:00) and the period of declining wind speed (14:00–16:05) (Table 3, Fig. 5).

The average rate of erosion (calculated over the initial period of 150 min of steady wind) increased eightfold from 0.1 mm min^{-1} at the dune base to 0.8 mm min^{-1} mid-way up the slope, beyond which the erosion rate decreased to zero, changing to deposition of sand in the flattened crestal area. The rate of deposition at the brink was 1.24 mm min^{-1}, a total thickness of 25 cm of sand accumulating at the brink during the entire period of observation (270 min) (Fig. 5).

The split of the data set into two time periods shows that during the second half of the measuring session (14:00–16:05), when the wind speed was

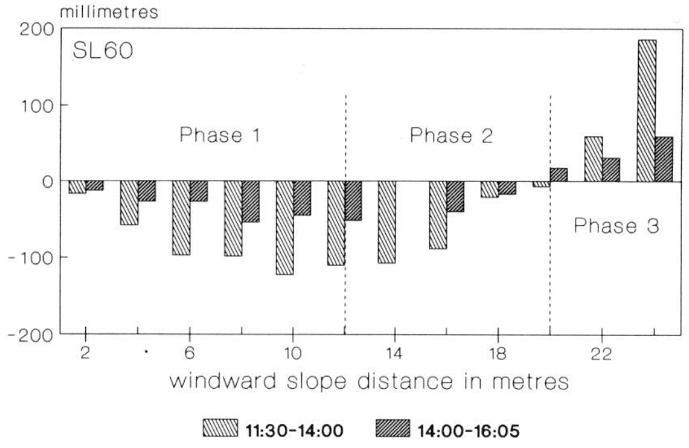

Fig. 5. Erosion pin data as measured along SL60 on the windward slope, 17 November 1988.

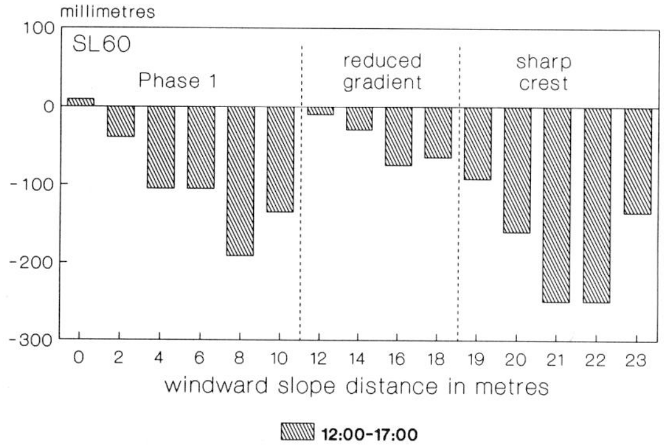

Fig. 6. Erosion pin data as measured along SL60 on the windward slope, 12:00, 7 December 1988.

decreasing, the nature of the interaction between windstream and dune surface did not change, but the rate of erosion was less; by estimation, about half the previous rate. Deposition continued in the crestal area during the period of decreasing wind speed.

At the start of the measurement session, the smoothly curved windward slope showed a gradual decrease in gradient from a value of 24° close to the dune base to a value of 20.5° 12 m upslope (Fig. 7, Table 3). Higher up the slope, the gradient decreased more rapidly by 3–4° per 2 m slope distance between 14/SL60 and 20/SL60 as the dune rounded towards the crest. The gradient decreased abruptly from 8.8° to 1.5° at the crest. The rounded shape of the

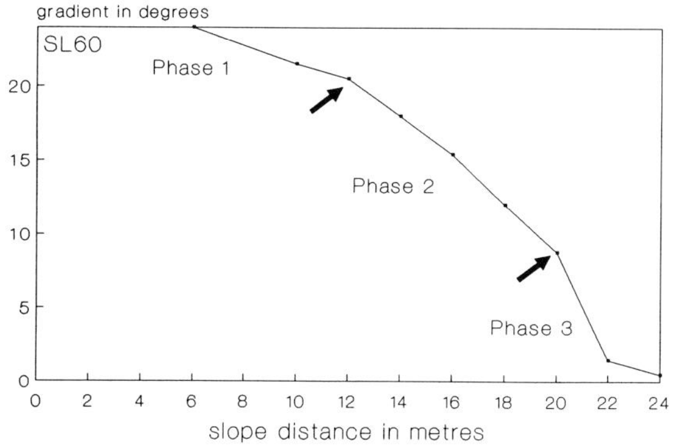

Fig. 7. Gradient measurements made at the erosion pins, 11:30, 17 November 1988. Arrows indicate inflection points.

Table 3. Erosion pin and gradient data monitored on the windward slope of SL60, 17 November 1988: negative values indicate erosion; positive values indicate deposition

Slope distance (m)/SL60	Sed. flux (mm) 11:30–14:00	Slope (degrees) 11:30	Sed. flux (mm) 14:00–16:05	Slope (degrees) 16:05
2	−17	—	−13	—
4	−58	—	−27	—
6	−97	24	−27	23.5
8	−98	—	−54	—
10	−122	21.5	−45	22
12	−110	20.5	−52	20.5
14	−107	18	—	19
16	−88	15.4	−40	15
18	−21	12	−17	13.5
20	−7	8.8	18	10.5
22	60	1.5	32	7
24	186	0.5	60	5

upper slope was the product of strong unidirectional easterly winds during the previous 2 days.

By the end of the measurement session, the gradient of the upper slope had steepened from less than 2° to 7° as a result of the accretion of sand in the crestal region (Table 3).

7 December 1988

The average rate of erosion (calculated over a period of 300 min of fairly steady wind speed) increased from a rate of 0.03 mm min^{-1} at 0/SL60 (base of the dune) to 0.6 mm min^{-1} at 8/SL60 (lower mid-slope). Erosion dropped off rapidly at 12/SL60 to 0.03 mm min^{-1} and then increased again towards the crest where 25 cm of sand was eroded during the entire 5 h period of observation; this represents an average rate of 0.83 mm min^{-1} (Fig. 6).

Whereas the windward slope appeared superficially to be a regular inclined surface sloping up at about 22°, terminating in a sharply triangular crest, a significant reduction in gradient was present in the upper mid-slope (Fig. 8). The crestal zone also showed a reduction in gradient, but not nearly as marked as was the case 3 weeks before. The general nature of the windward slope resulted from the southwesterlies which blew for 4 days about a week previously.

By the end of the measurement session the steepened gradient in the crestal region had decreased by several degrees, corresponding to the erosion in this region (Table 4).

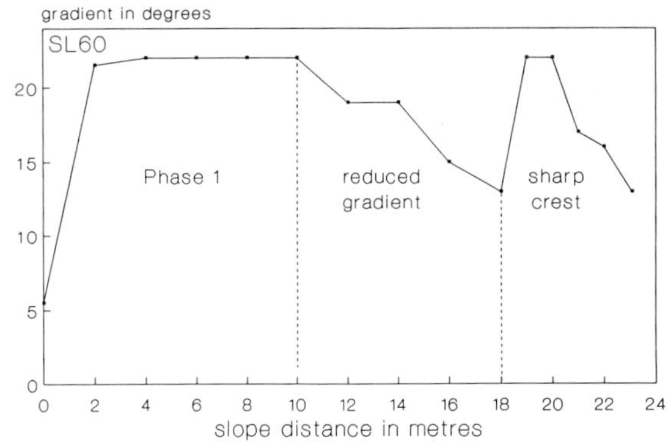

Fig. 8. Gradient measurements made at the erosion pins, 7 December 1988.

Table 4. Erosion pin and gradient data monitored on the windward slope of SL60, 7 December 1988: negative values indicate erosion; positive values indicate deposition

Slope distance (m)/SL60	Sed. flux (mm) 12:00–17:00	Slope (degrees) 12:00	Slope (degrees) 17:00
0	9	5.5	—
2	−40	21.5	19
4	−105	22	22
6	−105	22	22
8	−191	22	—
10	−135	22	—
12	−10	19	19.5
14	−30	19	16
16	−75	15	17
18	−65	13	15.5
19	−92	22	16.5
20	−160	22	15
21	−250	17	18
22	−250	16	15
23	−135	13	14

INTERPRETATION

The data set reported here provides insight into the inter-relationship between airflow response, local gradient measurements, change in local gradient and sand flux.

Airflow compression on 17 November resulted in the rapid acceleration of the wind speed from the base of the dune to 12/SL60 at 6 cm above the surface, with a corresponding increase in shear velocity. This is the region of maximum acceleration of the airflow (wind speed increased by 40% relative to the crestal wind speed). The progressive increase in erosion over this distance reflects a systematic and rapid increase in wind shear stress close to the surface. The decrease in the gradient of the slope was gradual, but gradient values were high compared to those of a conventional stoss slope.

Higher up the slope acceleration of the airflow closer to the surface diminished in response to a more rapid change of gradient as the dune rounded towards the crest. Reduction of compression resulted in lower shear stress close to the surface, which is confirmed by the decrease in erosion between 14/SL60 and 18/SL60.

The notable decrease in upslope gradient from 8.8° at 20/SL60 to 1.5° at 22/SL60 resulted in a marked reduction in wind speed between these two stations. The divergence of airflow and reduced shear stress caused the onset of deposition of sand at the crest. Thus deposition at the crest can be related to change of gradient rather than absolute gradient as suggested by Bagnold's formula, which predicts the minimum sand removal in regions of low gradient.

On 7 December the acceleration of the airflow against the lower slope of section line SL60 (which had a constant gradient of 22°) was reflected in the erosion trend; this implies a sufficient increase in shear stress to progressively increase the rate of erosion on the lower slope. As on 17 November, the gradient of the lower slope was steeper than that of a stoss slope produced under unidirectional conditions.

The abrupt decrease in the erosion rate at 12/SL60 corresponds to a lack of acceleration of the flow mid-slope. This could be attributed to divergence of the flow and incipient separation over the change in gradient at 12/SL60. The low rate of erosion between 14/SL60 and 18/SL60 could also be attributed to the continued reduction in gradient.

The normalized wind speed increased by 16% from 17/SL60 to 21/SL60 at the crest due to renewed compression of airflow against the steeper crest, which resulted in increased erosion towards the crest. The low rate of erosion at 18/SL60 could be ascribed to a sheltering effect induced by the abrupt steepening of the slope downwind at 19/SL60.

Comparison of the gradient values with the erosion pin data of 7 December shows that the maximum rate of erosion is associated with areas of maximum gradient (as predicted by Bagnold's formula), but there is little correlation between the degree of steepness and the amount of sand removal mid-slope. Here the non-uniform shape of the dune plus surface irregularities effect an airflow response which cannot be predicted by a formula based on a dune surface having an aerodynamic gradient.

SUMMARY AND CONCLUSIONS

On both occasions the windward face was undergoing reversal, so that the airflow response was constantly adjusting to the changing shape of the dune. Results show that variation in sediment flux can be related not only to absolute gradient but also to change in gradient. In addition, the curvature of the slope on 17 November was such that we recognize three different phases in the erosional response of the dune, giving an insight into the self-regulatory relationship between form and flow. The phases are determined by a change in the rate of change of gradient, and can be identified as inflection points in Fig. 7. Phase 1 is a region of flow acceleration corresponding to the area of maximum compression on the lower slope. The rate of erosion increases over this distance (Fig. 5) and the rate of change of gradient can be considered constant. Phase 2 is associated with an accelerated rate of change of gradient on the middle to upper slope. Both the airflow and the rate of erosion decrease over this distance. Phase 3 occurs in the flattened crestal area where flow divergence causes the airflow to lose contact with the surface. On 17 November the change in gradient was so marked in the crestal region that deposition occurred.

Although three phase types can be identified in the erosional response of the dune on 7 December, when the windward slope had a steep crestal profile, the relationship does not appear to be as well defined as for the data recorded on 17 November. The phase of acceleration on the lower slope correlates with Phase 1 of the model described above, but the erosional pattern mid-slope, coupled with the almost constant wind speed measured in this region, suggests a disruption in the airflow rather than controlled deceleration of the airflow, associated with Phase 2 of the proposed model. The development of a steep crest resulted in airflow acceleration and increased erosion in the crestal region. The rounding of the crest towards the brink corresponds to the reduction in gradient associated with Phase 2. Phase 3 was not attained on 7 December.

The distribution of wind shear stress on the dune surface is critical in determining the sediment flux over the dune surface, but values of shear stress will vary with the changing shape of the dune. The extent of shearing depends principally on the wind speed; there will be an increased wind speed gradient with a higher wind velocity. However, shear stress will be enhanced by increased compression which is related directly to the inclination of the surface. Wind direction is relevant in that the gradient of the surface as 'seen' by the wind alters with the approach angle of the wind. Maximum compression will occur when the wind approaches the windward face of the dune at 90° to the contour lines.

Bagnold's tangent relationship is valid for a slope with an aerodynamic gradient, such that the curvature of the slope is preserved with the advance of the dune, but it does not take into account surface irregularities even when the slope is in approximate equilibrium with the prevailing wind. Our results confirm that maximum shear stress does occur on the steepest slope where maximum compression occurs. However, airflow response appears to be very sensitive to changes in gradient, and surface shear stress is affected by surface irregularities.

Our study demonstrates how erosion pins act as a rapid and sensitive measure of the response of the dune surface to changes in shear stress distribution. In contrast, wind speed measured at 140 cm above the surface does not reflect subtle changes of shear stress at the surface. The two data sets reveal that a totally different erosional response was evoked under very similar wind conditions for what appeared superficially to be the same dune surface on two separate occasions.

The height of 6 cm above the surface for the lowest anemometers was an arbitrary height, chosen as a result of the limitations of our equipment. However, wind speed measured at 6 cm above the surface at five locations on the windward slope seems to register successfully general changes in shear stress which correspond to those recorded by the erosion pins.

Accurate and reliable measurement of airflow is particularly difficult in the field when attempting to work close to the dune surface. A more accurate determination of shear stress could be achieved with an array of pitot tubes within 20 cm of the dune surface; we have started to build suitable devices to pursue this work. It is, however, the erosion pin data which yield the finer resolution of shear stress changes on the dune surface, and in future it is possible that such simply obtained data could be used to interpolate shear stress where gradient, grain size and other factors relating to critical shear stress are known.

ACKNOWLEDGEMENTS

The Department of Environment Affairs, South Africa, funded the project through the Institute for Coastal Research, University of Port Elizabeth. Mr A.T. Sayers, Department of Mechanical Engineering, University of Cape Town, enabled the calibration of the anemometers. Werner Illenberger readily supplied the technical expertise to get the anemometer systems up and running, for which we are most grateful.

REFERENCES

Bagnold, R.A. (1941) *The Physics of Blown Sand and Desert Dunes*. Methuen, London, 265 pp.

Hesp, P.A., Illenberger, W.K., Rust, I.C., McLachlan, A. & Hyde, R. (1989) Some aspects of transgressive dunefield and transverse dune geomorphology and dynamics, south coast, South Africa. *Z. Geomorph. NF* **73**, 111–123.

Howard, A.D., Morton, J.B., Gad-El-Hak, M. & Pierce, D.B. (1978) Sand transport model of barchan dune equilibrium. *Sedimentology* **25**, 307–338.

Illenberger, W.K. & Rust, I.C. (1988) A sand budget for the Alexandria coastal dunefield, South Africa. *Sedimentology* **35**, 513–521.

Lai, R.J. & Wu, J. (1978) Wind erosion and deposition along a coastal sand dune. *Sea Grant College Program, University of Delaware, DEL-SG-10-78*, 26 pp.

Lancaster, N. (1985) Variations in wind velocity and sand transport on the windward flanks of desert sand dunes. *Sedimentology* **32**, 581–593.

Livingstone, I. (1986) Geomorphological significance of wind flow patterns over a Namib linear dune. In: *Aeolian Geomorphology* (Ed. Nickling, W.G.) pp. 97–112. Allen & Unwin, Boston.

Mulligan, K.R. (1988) Velocity profiles measured on the windward slope of a transverse dune. *Earth Surf. Proc. Landf.* **13**, 573–582.

Tsoar, H. (1983) Dynamic processes acting on a longitudinal sand dune. *Sedimentology* **30**, 567–578.

Watson, A. (1987) Variations in wind velocity and sand transport on the windward flank of desert sand dunes. Discussion. *Sedimentology* **34**, 511–520.

Spec. Publs Int. Ass. Sediment. (1993) **16**, 23–44

Late Quaternary development of coastal parabolic megadune complexes in northeastern Australia

K. PYE

Postgraduate Research Institute for Sedimentology, University of Reading, Whiteknights, Reading RG6 2AB, UK

ABSTRACT

Large coastal dunefields on the coast of northeastern Australia display stratigraphic and morphological evidence of a long period of sand accumulation during the Quaternary. Many of the older dune units are highly weathered and deeply podzolized, although the pattern of post-depositional alteration varies significantly in response to hydrological factors. Drilling has shown that in some areas aeolian sands extend to a depth of more than 40 m below present sea level, while elsewhere dunes have climbed over bedrock hills and older dunes to attain a height of more than 300 m above sea level. Insufficient evidence is presently available to be certain about the number, timing, extent and causes of aeolian episodes in different areas. However, the limited radiocarbon, thermoluminescence (TL) and uranium-series age data available suggest that some aeolian activity occurred during both high and lower stands of sea level. There is good evidence that significant transgressive dune building activity was associated with the early Holocene marine transgression, when coastal sediments were partially reworked landwards. After sea level attained approximately its present position around 6000 BP, shoreline erosion ceased and was replaced by progradation in many areas, leading to a reduction in sand supply from the beach and widespread dune stabilization. In northern Queensland stabilization may also have been encouraged by wetter conditions in the mid-Holocene. In some areas inland dunes have been reactivated and others initiated during the later Holocene following natural and anthropogenic disturbance to the vegetation.

INTRODUCTION

Despite more than a century of research, the factors which determine the morphology of aeolian sand accumulations remain poorly understood. In part this is because there have been few field studies of windflow and sediment transport, and partly it reflects the difficulty in extrapolating from small-scale and short-term studies to the scale of whole dunes and dunefields which may take decades or centuries to develop.

In broad terms, the form and scale of aeolian sand accumulations are likely to be determined by seven main factors:

1 sand availability;
2 grain size distribution;
3 wind velocity distribution and directional variability;
4 vegetation cover and growth characteristics;
5 the nature of the surrounding and underlying topography;
6 climatic and sea level changes; and
7 long-term patterns of tectonic uplift and subsidence.

Many of these factors are inter-dependent and changes in one may bring about a complex, frequently non-linear, morphological response. Despite the possible complexities, however, the evidence now available indicates that large barchans, transverse dunes and dome dunes are the predominant forms which develop in arid areas with a highly unidirectional wind regime. Where there is a significant cross-wind, these forms may display asymmetry, reversing behaviour, superimposed bedforms with a different orientation or some degree of sideways movement (see Pye & Tsoar, 1990, for a more detailed discussion). In markedly acute bidirectional wind regimes, linear dunes of the seif type are most likely to develop, while in obtuse bimodal and multimodal wind regimes, reversing, transverse and star dunes are the most common forms. In humid areas with relatively dense vegetation cover, formation of elongate parabolic dunes is favoured by a unidirectional effective wind regime, while in areas of bimodal or complex wind regimes parabolic dunes are likely to have a low length:width ratio and may be asymmetrical or digitate in form. Contrary to Wilson's (1972) suggestion, grain size does not appear to exercise a major control on the morphology and size of dunes, with the exception of zibar formation. Sand availability exercises some control on dune morphology, but does not always show a direct relationship to the morphology, size and spacing of individual dunes within an area. For example, although barchans tend to form in unvegetated areas when sand supply is relatively restricted, and transverse ridges dominate when sand is relatively abundant, individual barchans can attain megadune proportions even when sand is scarce, as indicated by the equivalent sand thickness (EST) in an area. In many sand seas the largest, most widely spaced dunes occur at the greatest distance from the source due to growth of the bedforms as they move downwind, or to long-term accumulation of sand as successive generations of dunes become superimposed within a topographically or climatically controlled sand sink.

This paper examines some of the factors which have influenced the development, form, orientation and structure of parabolic dunefields along the coast of northeast Australia. Large coastal sand masses have formed in southern Queensland as a result of long-term accumulation of sand transported northwards along the inner continental shelf from New South Wales. Most of the dunes have formed under the influence of predominant southeasterly winds, although there are significant regional and local variations in wind regime. Most of the dunes in southern Queensland have a relatively low length:width ratio, and some are transitional between parabolic and transverse forms, reflecting a bimodal to polymodal wind regime. By contrast, parabolic dunes in central and northern Queensland are typically much more elongate, reflecting a more unidirectional wind regime. Many of the Queensland dunes are more than 200 m high and several kilometres in length. The exceptional size of these megadunes reflects the combined effects of high wind energy, abundant sand supply provided by landward reworking of sand during periods of changing sea level, and the resistance created by the regional vegetation to downwind dune migration, leading to stacking and superimposition of several generations of dunes during the later Quaternary.

STUDY AREA

Large coastal dunefields occur at several locations in northeast Australia, most notably on the east coast of Cape York Peninsula and in southern Queensland. Smaller dunefields occur on the coast and offshore islands of central Queensland (Pye, 1983a). No large part of the coast is entirely without dunes, although they are poorly developed or localized on sections of coast which are sheltered from, or are aligned parallel to, the prevailing southeasterly trade winds. These include the western shores of Broad Sound, the Proserpine to Townsville sector, Princess Charlotte Bay and the western coast of Cape York Peninsula (Fig. 1).

The accumulation of coastal sediments during the Quaternary has been strongly influenced by tectonic factors which have determined the coastal orientation, physiography and lithological outcrop pattern. The entire eastern continental shelf has experienced slow subsidence since the late Tertiary, associated with block faulting along a predominant north-northwest–south-southeast axis, while the eastern part of the mainland has experienced varying degrees of uplift (Maxwell, 1968).

The shelf shows marked contrasts to the north and south of Sandy Cape on Fraser Island. To the south of this point the shelf has a maximum width of 40 km and is relatively steep with few offshore islands. The wave climate is dominated by long-period ocean swell. To the north, the shelf widens to a maximum of 250 km between Rockhampton and Mackay, is characterized by a gentler gradient, and is broken by numerous high islands of continental rock and coral reefs (Bird, 1971a). Average water depth exceeds 60 m around Mackay, falling to 30 m north of Townsville, reflecting greater subsidence of the shelf towards the south. Wave energy on the semi-protected shelf of central and northern Queensland is dominated by short-period wind waves generated over moderate fetches. As in southern Queensland, the predominant wave approach direction is southeasterly, resulting in a net northerly drift along much of the coast, but both longshore and cross-shore sediment transport are

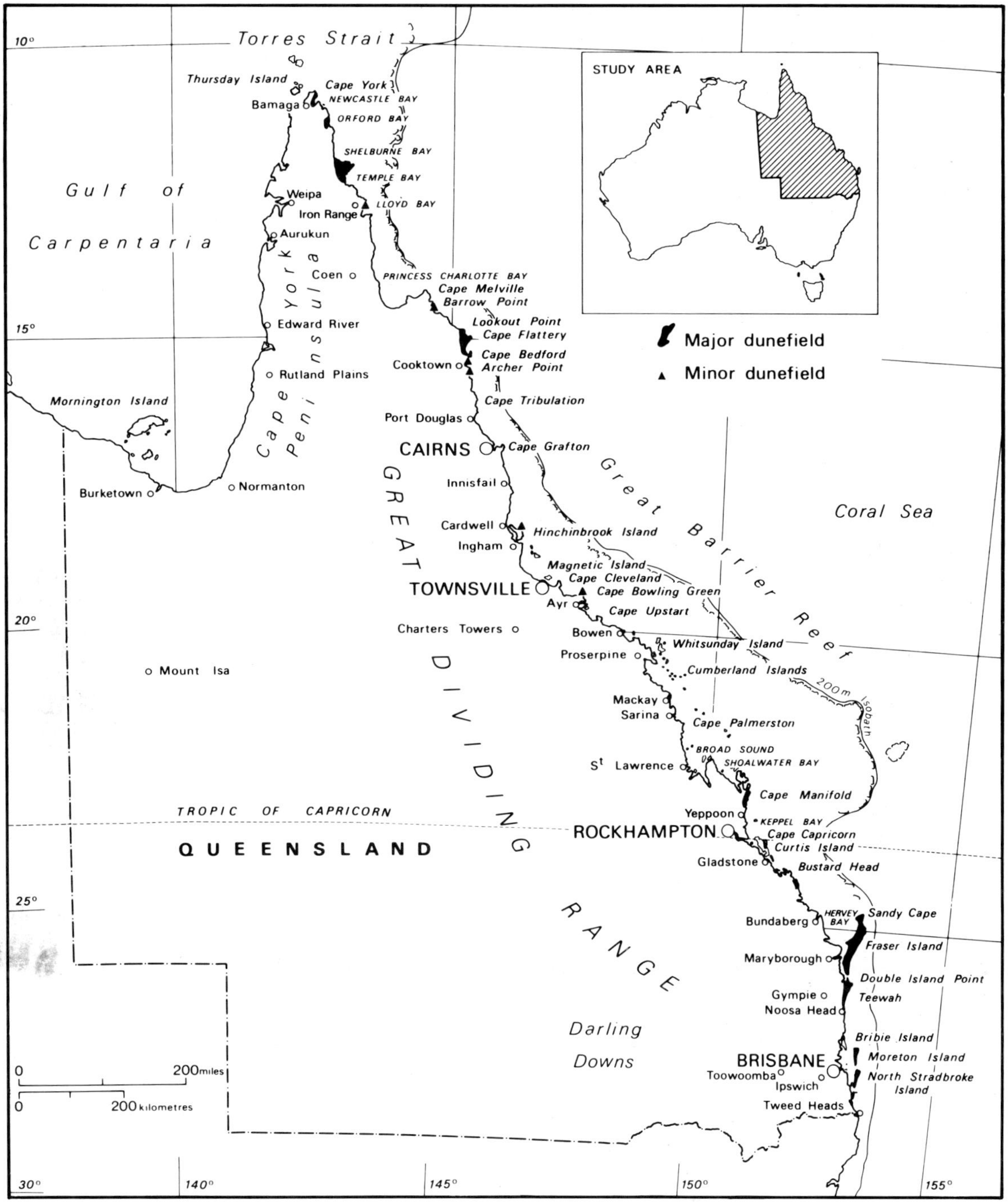

Fig. 1. Location of major coastal dunefields in northeastern Australia.

more restricted within the Great Barrier Reef Province (Maxwell, 1968).

Pye (1983a) divided the Queensland coast into four sections on the basis of the degree of dune development:

1 the west coast of Cape York Peninsula and the Gulf of Carpentaria lowlands, characterized by extensive beach ridges and chenier plains with very small-scale dune development (e.g. Rhodes, 1982);

2 the east coast of Cape York Peninsula, which contains a number of very large quartz dune complexes, notably along the shores of Orford Bay, Shelburne Bay, Temple Bay and between Cape Bedford and Cape Flattery (Bird, 1965; Pye, 1982a, 1983b);

3 northern and central Queensland, where coastal dune complexes are localized and generally on a smaller scale than in eastern Cape York and southern Queensland; and

4 southern Queensland, which includes the large sand dune islands of Fraser Island, Moreton Island and North Stradbroke Island, together with the large mainland dune complex at Cooloola (Coaldrake, 1962; Ward, 1977; Thompson, 1983).

Mineralogical evidence suggests that much of the sand comprising the southern Queensland sand masses is derived from river catchments in northern New South Wales, and the area appears to have acted as a long-term sink for sediment which drifted northwards along the continental shelf (Roy & Thom, 1981). A significant proportion of the sand is, however, derived from the Moreton and Maryborough Basins of southern Queensland. Virtually all of the coastal dune sand within the Great Barrier Reef Province is siliceous, and is also derived from terrigenous sources in Queensland. The latter include catchments with major outcrops of granitic, metamorphic and arenaceous sedimentary rocks. Extensive areas of Cape York Peninsula are covered by blankets of residual and alluvial siliceous sand, in places several metres thick, which is transported to the coast by rivers or reworked *in situ* by transgressive dunes (Smart *et al.*, 1974; Wright, 1976; Willmott & Powell, 1977). Carbonate dunes are extremely limited in extent, occurring only in small isolated bays and on islands which receive little terrigenous sediment (Bird & Hopley, 1969; Bird, 1971b).

The climate of the Queensland coast ranges from seasonally humid tropical in the north to humid subtropical in the south. The prevailing winds are the southeasterly Coral Sea Trades which blow particularly strongly in northern and central Queensland during the dry season (April–October) (Fig. 2). During the wet season they are periodically replaced in northern Queensland by a weak northwesterly airflow from the Gulf of Carpentaria (Gentilli, 1971). At this time strong winds of short duration may blow from any direction associated with the passage of tropical cyclones (Lourensz, 1977).

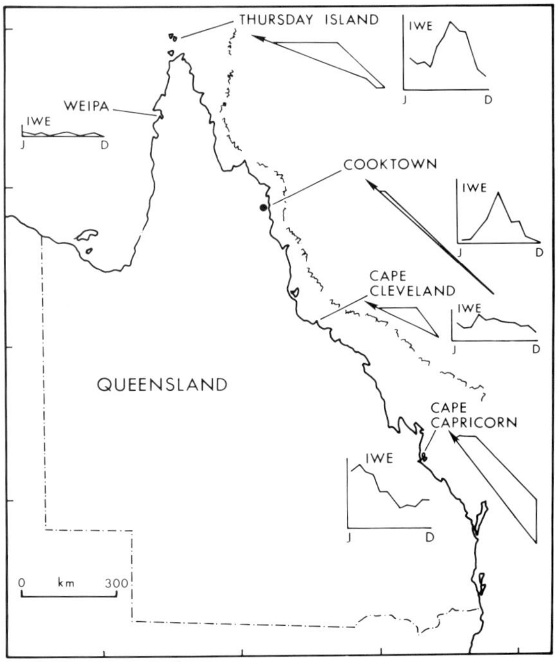

Fig. 2. Resultant wind vectors for effective sand-moving winds at four Queensland coastal stations, calculated using data supplied by the Bureau of Meteorology. The index of wind energy (IWE) is calculated after Pye (1985).

Queensland as a whole is comparable with other trade wind coasts such as southeastern Brazil and Mozambique in terms of its total wind energy, but there are significant variations between recording stations. Stations lying to the south of the Tropic of Capricorn show a slightly higher mean wind energy than do stations to the north of the Tropic (Pye, 1985). Wind energy at the coast is relatively low between Proserpine and Ingham, where a high proportion of winds blow almost parallel to the coast. Many of the mainland beaches along this section of coast are also sheltered by offshore islands, and the presence of high coastal ranges forces the winds to rise, creating relatively sheltered conditions at their foot (Pye, 1983c). Dunes on this section of coast are generally restricted to bays on the south side of major headlands and estuarine barrier spits with a southeasterly orientation (Figs 3 and 4).

There are also important latitudinal variations in the seasonal distribution of wind energy. At Thursday Island in Torres Strait, >74% of the total annual wind energy occurs during the dry season (May–November), but this percentage declines southwards to 68% at Low Isles (near Port Douglas), 61% at Cape Cleveland, 50% at Pine Islet (near Mackay) and 38% at Caloundra, north of Brisbane (Pye, 1985).

The most unidirectional winds are found on the east coast of Cape York, north of Port Douglas. At Cooktown, >99% of the effective sand-transporting winds blow from the southeast, where the vector resultant is N134°E (Fig. 2). At Thursday Island, where there is a larger easterly wind component (Swan, 1979, 1981), the vector resultant is N110°E. In southern Queensland, winds from the south are more important than further north, and there is greater overall directional variability.

The significance of local variations in wind regime caused by coastal orientation and the nature of surrounding topography is clearly demonstrated within individual embayments. At Ramsay Bay on Hinchinbrook Island, for example, a 10 km long tombolo barrier is capped by parabolic dunes which increase in size and complexity towards the more exposed northern end of the embayment (Pye,

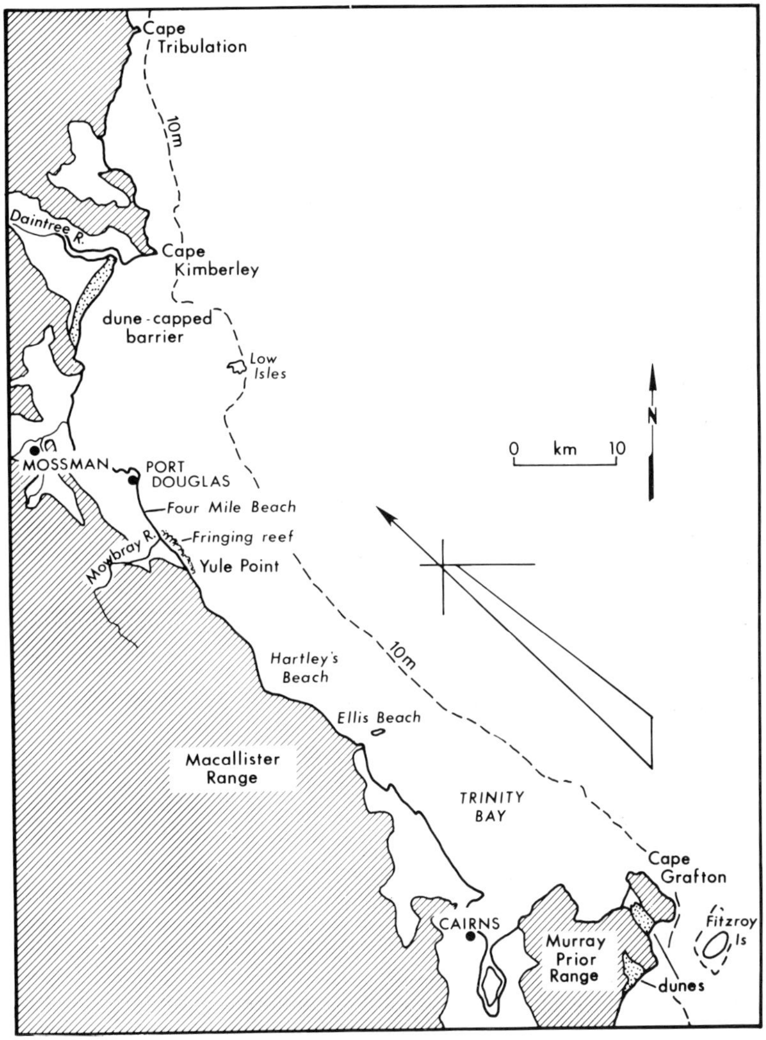

Fig. 3. The coast between Cape Grafton and Cape Tribulation, northern Queensland, showing the location of small dunefields and the Low Isles wind resultant.

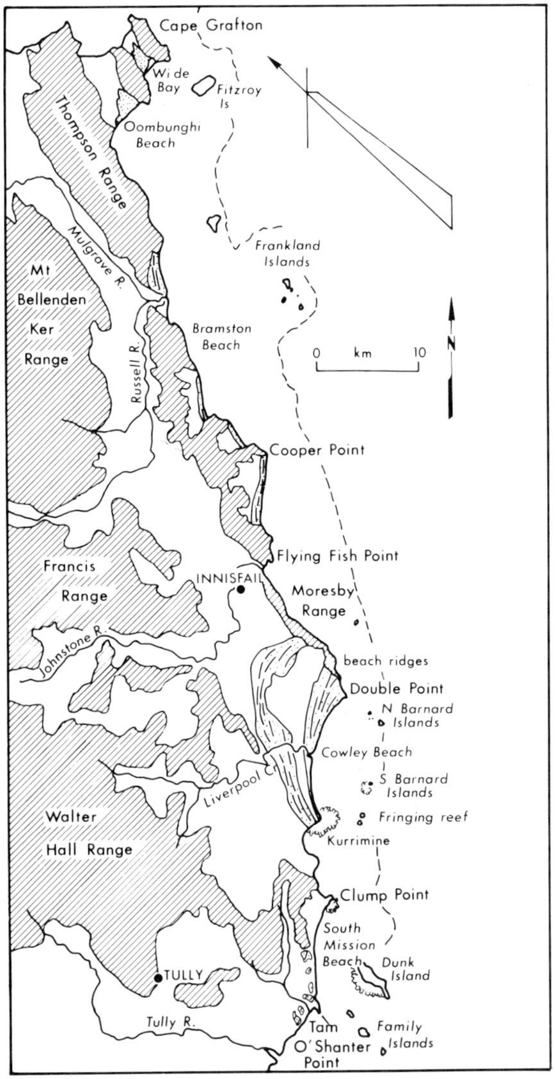

Fig. 4. The perennially wet tropical sector of the northern Queensland coast between Dunk Island and Cape Grafton. Parabolic dunes occur only on the exposed beaches at Wide Bay and Oombunghi Beach; low foredunes and aeolian cappings superimposed on beach ridges are present at Cowley Beach. The Fitzroy Island wind resultant is also shown.

1982b). Most of the very large dune systems on the coast occur where the northerly drift of nearshore sand has been impeded by a prominent bedrock headland, creating a closed 'sediment compartment' (cf. Davies, 1974) or 'receptive shore' (cf. Cooper, 1958). On the leeward side of such headlands, the sediments are frequently poorly sorted; mudflats, mangroves and chenier ridges are the typical depositional forms (Bird & Hopley, 1969; Frankel, 1974; Belperio, 1983; Chappell *et al.*, 1983; Chappell & Grindrod, 1984).

Present-day rates of sediment supply to the nearshore zone are high near the mouths of major rivers such as the Burdekin, Pioneer and Normanby (Belperio, 1983; Pye, 1983a). Although dunes represent an important component of these active depositional systems, they are less well developed than in other areas which today appear to receive little, if any, sand from terrigenous or continental shelf sources. The majority of the major dune systems appear to be largely relict features which are presently undergoing varying degrees of *in situ* reworking.

DUNEFIELD MORPHOLOGY

Dune formations on the Queensland coast occur in a number of different settings and consequently display differing morphologies and stratigraphies. The largest transgressive dune complexes correspond to types E and F shown in Fig. 5. In the case of type E, represented by the Cape Manifold (Fig. 6) and Cape Flattery dunefields (Fig. 7), transgressive dunes have climbed over coastal hills, alluvial plains and older generations of dunes to produce a stacked sequence of aeolian sediments. The truncated trailing arms of some of these dunes suggest that they originated some distance seaward of the present shoreline, possibly at a time of lower sea level. In several places, including Cooloola (Fig. 8) in southern Queensland, coastal erosion has exposed a thick sequence of highly weathered coloured sands, representing several generations of podzolized dunes extending back in age beyond the last inter-glacial period (Coaldrake, 1962; Thompson & Hubble, 1977; Ward *et al.*, 1979; Thompson, 1981, 1983).

Type F shown in Fig. 5 illustrates the structure typical of the southern Queensland sand islands, which consist of a stacked series of partially drowned transgressive dunes anchored by bedrock highs (Laycock, 1978). Initiation of the sand islands clearly began at a time when sea level was lower than at present (Ward, 1977), although some transgressive dunes were also active during periods of marine transgression and high stand. During the Holocene marine transgression the margins of the major sand islands, including Stradbroke Island

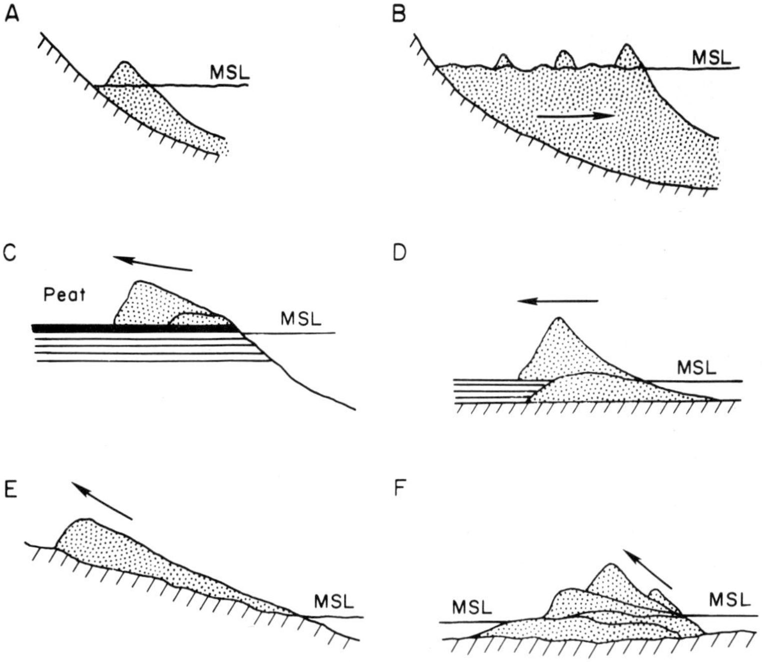

Fig. 5. Six contrasting styles of aeolian accumulation found on the Queensland coast. Type B represents the typical morphology found on prograding shores at the present day. Types C, D, E and F are typically associated with coastal erosion. (After Pye, 1983a.)

(Fig. 9) and Moreton Island, were trimmed by marine erosion (Ward, 1982), but transgressive dunes which originated at or near the eroding shoreline continued to move inland and climb over older, stabilized dunes. Sand eroded from exposed parts of the sand islands has, in some instances, been deposited in more sheltered areas as progradational beach ridge plains (e.g. Fig. 9).

PARABOLIC DUNE MORPHOLOGY

Dunes which are loosely described as 'parabolic' show a variety of forms in detail. A distinction can be made between *simple*, *compound* and *complex* parabolic forms, following the distinction made by McKee (1979). Simple parabolic dunes consist of a single dune form which contains three basic elements, a 'nose' or apical sand mound at the downwind end, two trailing 'arms' or ridges, and a deflation corridor or trough which is open at the upwind end. Simple parabolics may be further subclassified on the basis of their length:width ratio (Fig. 10A). Dunes with a length:width ratio of <0.4 are described here as *lunate*, those with a length:width ratio of 0.4–1.0 are described as *hemicyclic*, those with a length:width ratio of 1.0–3.0 are described as *lobate*, and those with a length:width ratio of >3.0 are referred to as *elongate*. A schematic model for the formation of elongate parabolic dunes from an isolated blowout is presented in Fig. 11.

Compound parabolic dunes consist of a number of adjoining or superimposed parabolic forms. Four end-member variants are shown in Fig. 10C: *nested*, *en-echelon*, *digitate* and *superimposed*. In reality, actual field examples are often transitional between these end-members.

Complex parabolics may consist of a wide variety of architectural elements in varying combinations (Fig. 10B). One of the most common types found in Queensland consists of a broadly parabolic-shaped sand ramp or sand sheet upon which are superimposed transverse ridges or small barchan forms. The margins of the transgressive sand sheet are often defined by a partially vegetated ridge. They are similar to the 'long-walled transgressive ridge' dunes described in New South Wales (Thom *et al.*, 1978) and a few resemble the 'precipitation ridges' described in Oregon by Cooper (1958). Some of the vegetated complex parabolic dunes found on the southern Queensland sand islands are exceptionally large, measuring more than 6 km in width and 10 km in length (Figs 8 and 9). These megadunes have clearly experienced a complex evolutionary history, involving several phases of

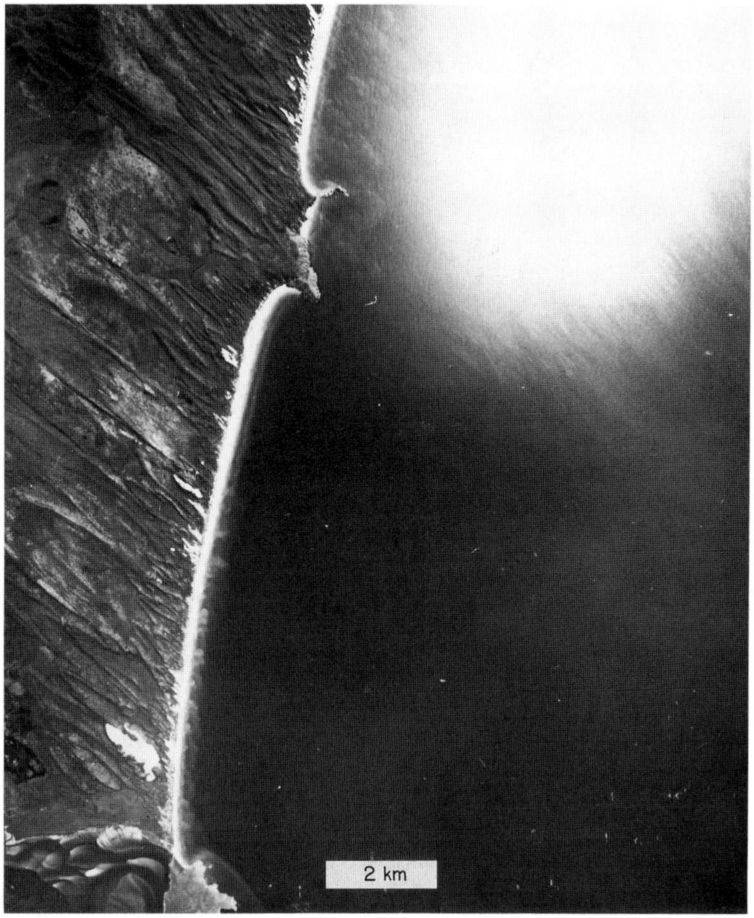

Fig. 6. Vertical air photograph of vegetated nested and superimposed elongate parabolic dunes south of Cape Manifold, central Queensland. (Reproduced by permission of the Department of National Mapping, Queanbeyan.)

aeolian stability, stabilization and partial reactivation or reworking by younger transgressive dunes.

The variation in parabolic dune morphology observed in different dunefields along the Queensland coast appears to be due both to variations in wind directional variability at the regional scale and to the influence of local topography which is responsible for local deviations in the wind regime (e.g. Fig. 9). The typical elongate parabolic dune forms in the Cape York Peninsula dunefields reflect the highly unidirectional nature of the regional wind regime. The more irregular plan form of the dunes on the southern Queensland sand islands appears to reflect a more variable regional wind regime and important local topographic influences. Although there are some differences in the nature of the dune vegetation in different parts of Queensland (Coaldrake, 1961; Lavarack & Stanton, 1977; Pye & Jackes, 1981), the sequence of plant communities is broadly similar and is likely to influence the dune morphology in similar ways.

Dune size is clearly related to sand availability. The megadunes which occur in both northern and southern Queensland have formed as a result of aeolian reworking of older dune units. Complex and compound forms have resulted where several generations of dunes have been initiated in relatively quick succession, often as a result of blowout development within otherwise stabilized dunes and sand sheets (e.g. Figs 12 and 13).

STRATIGRAPHY AND AGE STRUCTURE OF THE DUNEFIELDS

Cape Bedford–Cape Flattery dunefield

Drilling investigations at a number of locations

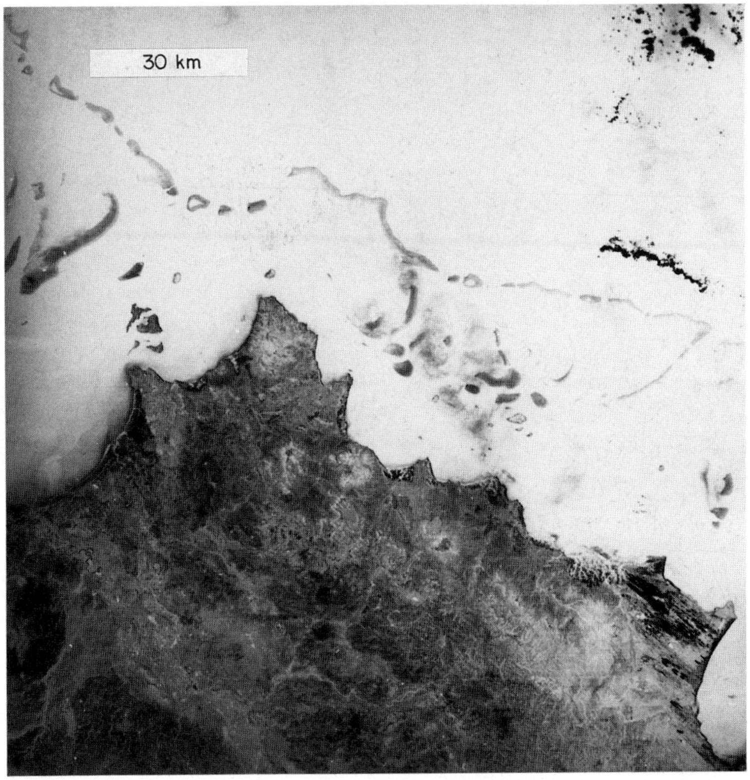

Fig. 7. Satellite photograph of the southeastern part of Cape York Peninsula, showing the Cape Bedford–Cape Flattery dune complex (black) on the lower right.

in Queensland (Thompson, 1981, 1983; Pye & Bowman, 1984; Pye & Rhodes, 1985) have confirmed the stratigraphic complexity of many of the dunefields. The older dune generations are typically highly weathered and show varying degrees of reddening and podzol soil development (Ward *et al.*, 1979; Pye, 1981, 1982c, 1983d; Thompson, 1983; Thompson & Bowman, 1984). Three contrasting stratigraphic examples from the Cape Bedford–Cape Flattery dunefield are shown in Fig. 14. Case A illustrates a relatively simple case of an active elongate parabolic dune composed of pure quartz sand which has transgressed over a vegetated aeolian sandplain surface. This surface is underlain by deeply podzolized sands characterized by a 3–4 m deep bleached A2 horizon and a 3–4 m thick humate stained Bh horizon. The bleached A2 horizon sands have been eroded by the wind in the deflation trough down to the level of the dry-season water table, and have been incorporated into the mobile dune.

Case B represents a more complex situation with two stacked generations of transgressive dunes overlying an old sandplain surface. The older of the two dune generations became stabilized and suffered podzolization before being buried and partially reworked during migration of the younger dune. During the period of stabilization a significant thickness of the older dune sand body became stained by ferric oxyhydroxides and some layers were cemented to form 'petroferric ironstone layers'. Subsequent deflation during the passage of the younger dune exposed the weathered sands at the surface of the deflation corridor, and the incorporation of ferruginized grains has given the active dune sands a pale orange colouration. Radiocarbon dating studies (Pye, 1981; Pye & Switsur, 1981) have shown that the podzolic reddening process at Cape Flattery is relatively rapid, taking only a few thousand years. Progressive podzolization over longer time periods leads to gradual removal of iron from the profile and accumulation of dark brown to black humic colloids within the zone of groundwater fluctuation (Pye, 1982c, 1983d; Thompson, 1983). Over very long time periods, the humic matter may itself be destroyed, leaving a residual sand cemented

Fig. 8. Vertical air photograph showing part of the Cooloola sand mass, southern Queensland. Note the variable orientation of the active sand blows, which reflects topographic influences on local wind patterns. (Reproduced by permission of the Department of National Mapping, Queanbeyan.)

only by kaolinite and gibbsite (Ward *et al.*, 1979; Pye, 1983c). During the process of weathering the sharpness of the forms becomes subdued by subaerial processes (Ward, 1977; Pye, 1983d; Thompson & Bowman, 1984).

Case C in Fig. 14 represents a situation involving three dune generations transgressing over a deeply podzolized sandplain surface. In this case, however, the migration of successive dunes has not involved significant reworking of the underlying material, leading to preservation of the former depositional surfaces within a stacked sedimentary sequence.

The age structure of the dunes on the Queensland coast has received considerable attention in the last few years (Ward, 1977; Grimes, 1979; Cook, 1986; Ward & Grimes, 1987; Coventry, 1988; Lees *et al.*, 1990). Pye & Switsur (1981) identified three generations of dunes at Cape Bedford and Cape Flattery based on radiocarbon dating of included wood and charcoal. The oldest dune unit, which is exposed in low coastal cliffs northwest of Cape Flattery (Fig. 13), is deeply podzolized and is partly cemented by humate. An *in situ* tree root from this formation gave a 'background' age of >48 000 ^{14}C yrs BP (Q-2085). Three samples of charcoal from the second dune generation identified (section located near the Cape Flattery Mining Camp), which is also podzolized and extensively reddened, gave ages of 7480 ± 75 (Q-2081), 7560 ± 90 (B-2494) and 8200 ± 85 ^{14}C yrs BP (Q-2082). The youngest generation of dunes, including those still active, partly overlies the two older generations (Fig. 12). Using measurements of the modern rate of dune movement and air photograph evidence, Pye (1982a) concluded that activation of many of these dunes has occurred within the last few hundred years.

Lees *et al.* (1990) examined a 40 m section with numerous palaeosols and six inter-bedded sand

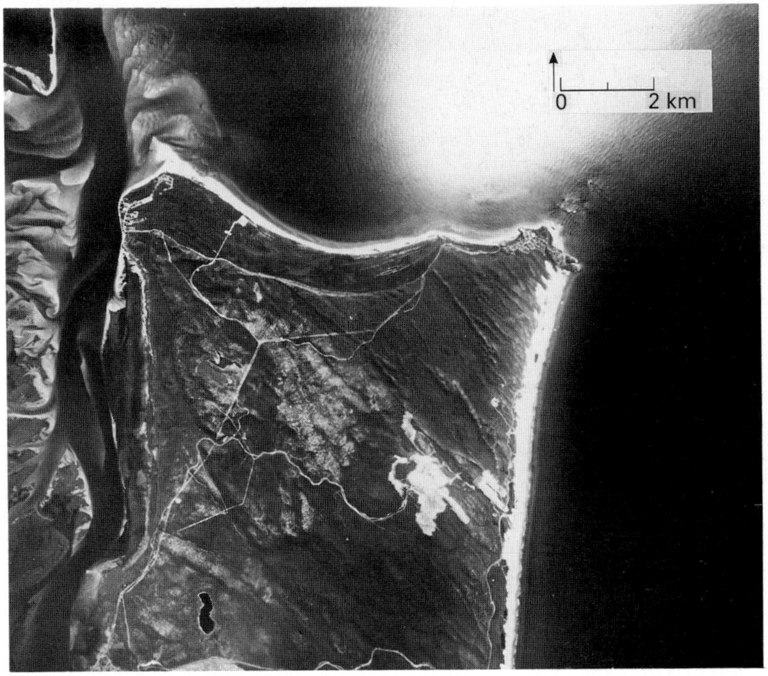

Fig. 9. Vertical air photograph of the northern part of North Stradbroke Island, showing stabilized parabolic megadunes which have been truncated by marine erosion, followed by a period of beach ridge plain construction. (Reproduced by permission of the Department of National Mapping, Queanbeyan.)

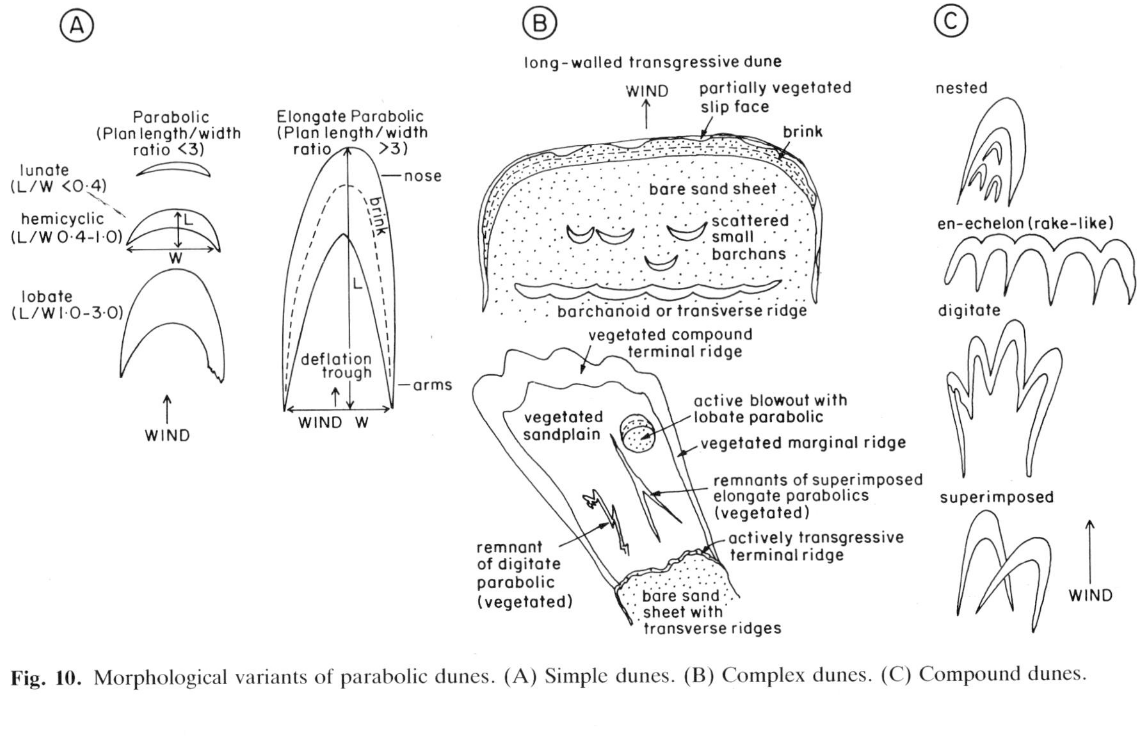

Fig. 10. Morphological variants of parabolic dunes. (A) Simple dunes. (B) Complex dunes. (C) Compound dunes.

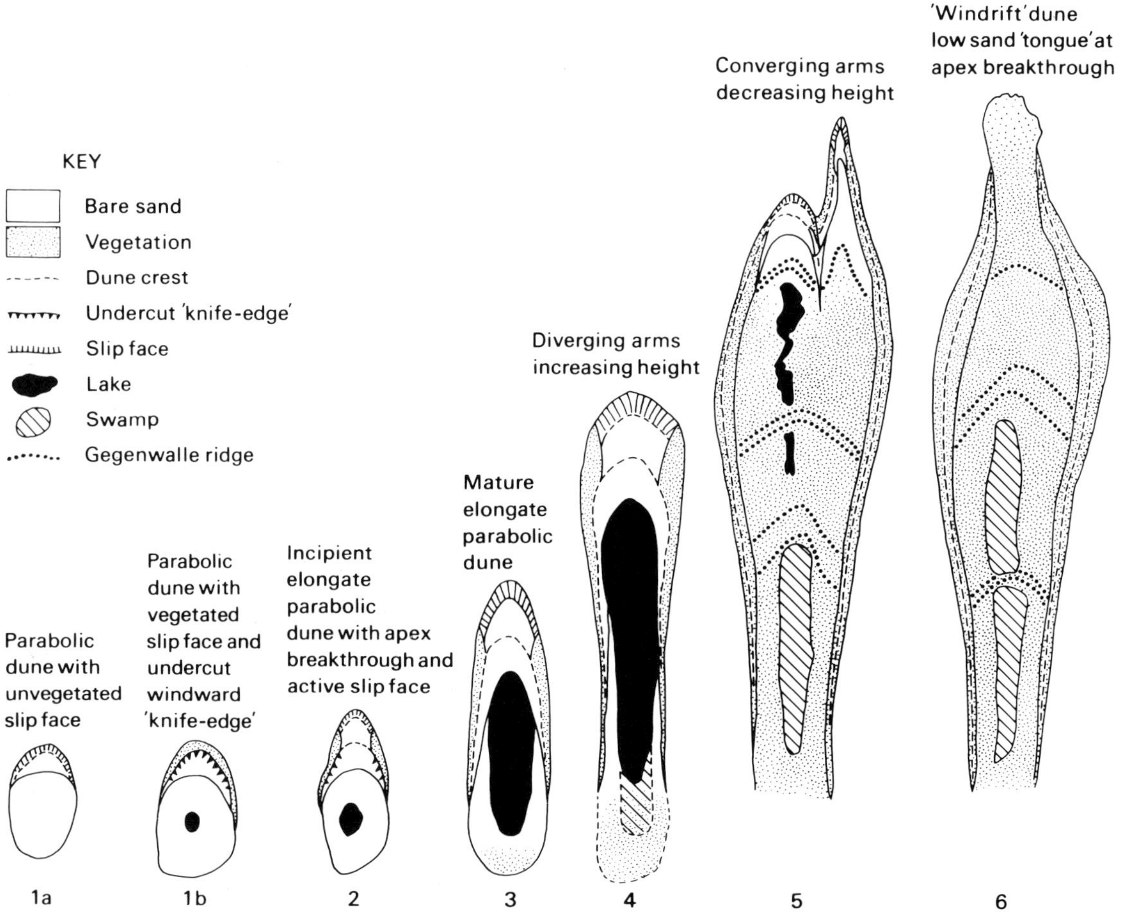

Fig. 11. Schematic model showing the evolutionary development of elongate parabolic dunes from a spot blowout at Cape Flattery. (After Pye, 1982a.)

layers exposed in a blowout near Red Hill, approximately 12 km southwest of the Cape Flattery settlement. The sand layers were designated A–F (from top to bottom) and samples collected for radiocarbon and thermoluminescence (TL) dating. Charcoal collected from a depth of 0.6 m at the base of Unit A, composed of pale yellowish sand, gave an age of 220 ± 70 ^{14}C BP (ANU 6171). A TL sample collected from Unit B at a depth of approximately 2 m gave an age of 2000 ± 1000 BP (Xian Au-03). A sample taken from Unit C at 10 m depth gave a TL age of 19 200 ± 3400 BP (Xian Au-18), a sample from Unit D at a depth of 20 m gave an age of 22 700 ± 2800 BP (Xian Au-19), and a sample from Unit E (separated from D by an erosional break) gave an age of 19 800 ± 2300 BP (Xian Au-04). Since these ages overlap at 1 standard deviation, no significant difference in the age of deposition can be inferred. Below this deposit, however, is a highly indurated, kaolinitic sand unit (Unit F) which gave a TL age of 171 000 ± 13 000 BP (Xian Au-050) (Lees *et al.*, 1990). In places Unit F has also been eroded, exposing an even more firmly indurated unit which could not be sampled for TL. A sample collected from an adjacent site at an elevation inferred to be stratigraphically lower than Unit F gave a minimum TL age of >66 000 yrs (α-2589).

The sequence investigated by Lees *et al.* (1990) is similar to that described briefly by Pye (1983d) at Elim Cliffs, northwest of Cape Bedford (Fig. 15). The sequence here consists of an uppermost unit of white to very pale yellow sands, associated with presently active dunes, overlying more than 60 m of partially reddened sands which contain four

Fig. 12. Vertical air photograph of the southern part of the Cape Flattery dunefield, showing currently active compound parabolic megadunes which are transgressing over an older sandplain surface composed of indurated dune sands. (Reproduced by permission of the Department of National Mapping, Queanbeyan.)

major erosional discontinuities (Fig. 16). Mid-way between the Elim Cliffs section and the McIvor River, two distinct, highly indurated and cross-bedded sandrock horizons are exposed just above the high-water mark. The uppermost unit, 2–3 m thick, is dark brown to yellowish brown, while the lower unit, at least 2 m thick, is paler in colour. These units are clearly of considerable age. Although they were grouped together within the Flattery Sand Unit by Pye & Switsur (1981), it is by no means certain that they are of the same age as the deeply podzolized and humate cemented sands exposed along the shores of Flattery Bay.

If the TL ages reported by Lees *et al.* (1990) are reliable, the evidence would suggest that transgressive dunes were active in the Cape Flattery area during the last glacial maximum when the sea level stood more than 100 m lower and the shoreline lay along the shelf edge on the outer side of the ribbon reefs. However, as pointed out by Pye (1984), the regional winds would have needed to have been much stronger to maintain dune movement at a distance of more than 40 km from the shore, since wind velocities today show a marked reduction only a few kilometres inland. During low sea level stands the continental shelf would have been characterized by highly irregular topography, emergent reefs and deeply incised fluvial channels (Maxwell, 1968). These would have further served to reduce wind velocities before they reached the present Cape Bedford–Cape Flattery area. For these reasons, Pye (1984) proposed that the older pre-Holocene dune units in the area are more likely to have been associated with earlier inter-glacial or inter-stadial high sea levels. Further dating information, including confirmation of the reliability of the TL ages, is required before this issue can be adequately resolved.

Potentially serious problems may arise in luminescence dating of sand grains collected from environments where podzolic weathering is intense, pedogenic translocation and horizon differentiation are major processes, and where the water content of the sediments varies significantly over seasonal and longer time scales. During podzolization, the feldspar and heavy mineral fractions are progressively attacked, and quartz grains themselves may suffer severe alteration (Little *et al.*, 1978; Pye, 1983e). These changes may affect both the internal and external dose rate received by quartz grains, and may affect the capacity of electron traps to retain stored energy. Annual dose rate determinations, measured either in the field or in the laboratory, are likely, in the case of weathered and leached podzolic A horizon sands, to underestimate

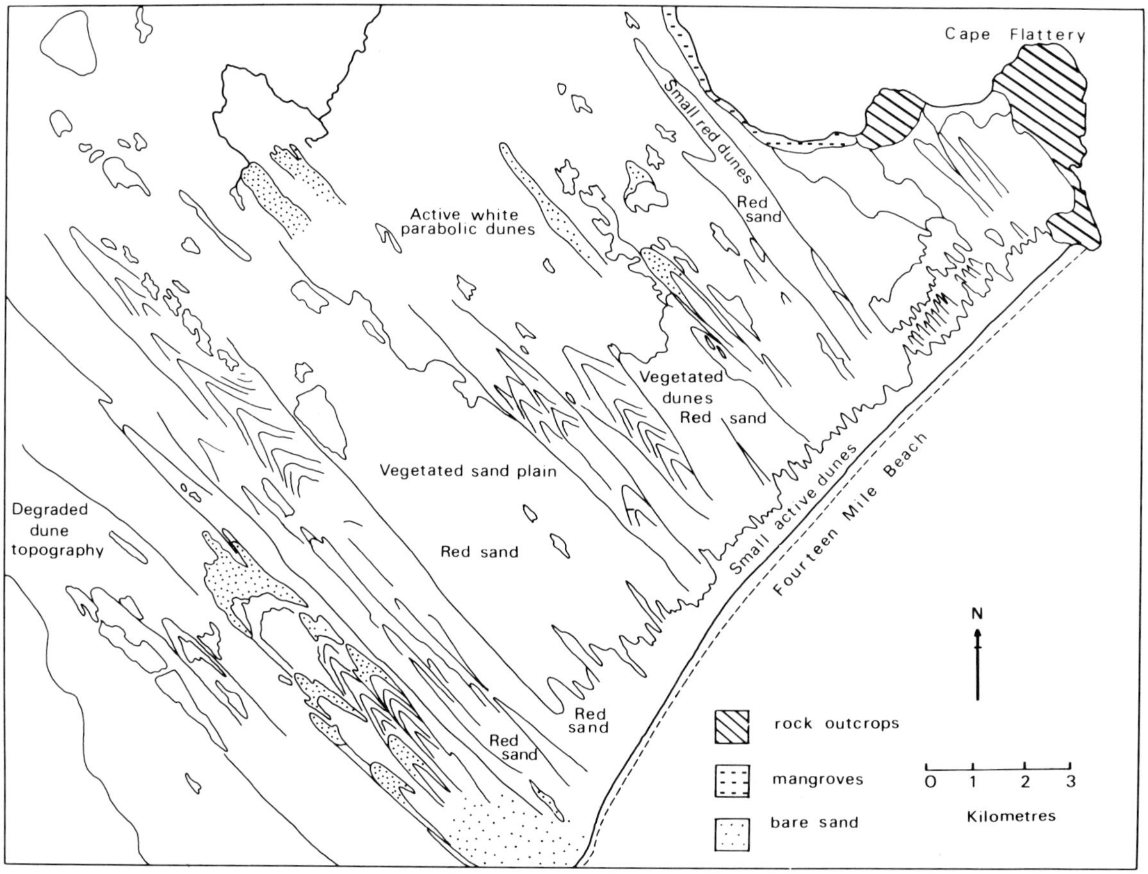

Fig. 13. Sketch showing the major morphostratigraphic features of the Cape Flattery dunefield, based on air photograph interpretation and field survey.

the average annual dose rate to which grains have been subjected, whilst in the case of some podzolic B horizon sands they may overestimate it due to long-term accumulation of U and Th. Reported TL age estimates which take insufficient account of the possible effects of weathering, translocation and water content changes may be in significant error.

Shelburne Bay dunefield

Lees *et al.* (1990) also reported TL results for samples collected from three sections in Shelburne Bay and one section at the northern margin of the Temple Bay dunefield. A TL date of 17 600 ± 1400 BP (α-2597) was obtained from the fine fraction (4–11 μm) of a podzolic B horizon at a depth of approximately 8 m in the first Red Cliffs section. This was interpreted by the authors to be a minimum age due to the possibility of overestimation of the average annual dose rate within the podzolic B horizon and mechanical translocation of fines through the sand profile, as discussed above. A similar age of 15 000 ± 1110 BP (α-2594), also interpreted as a minimum estimate, was obtained from the fine fraction of ferruginized sands with similar appearance near Conical Hill, northwest of Cape Grenville.

At the second Red Cliffs site, located 100 m away from the first, a TL age of 28 400 ± 1300 BP (Xian Au-07) was obtained from pale brown sand, overlain by white sand, at a depth of approximately 6 m. The brown sands are underlain by grey sand with brown pipes which yielded a TL age of 29 900 ± 2400 BP (Xian Au-09). At the base of this section is a humicrete containing detrital logs, the cellulose fraction of one of which gave a radiocarbon age of 29 740 ^{14}C yrs BP (ANU 4980).

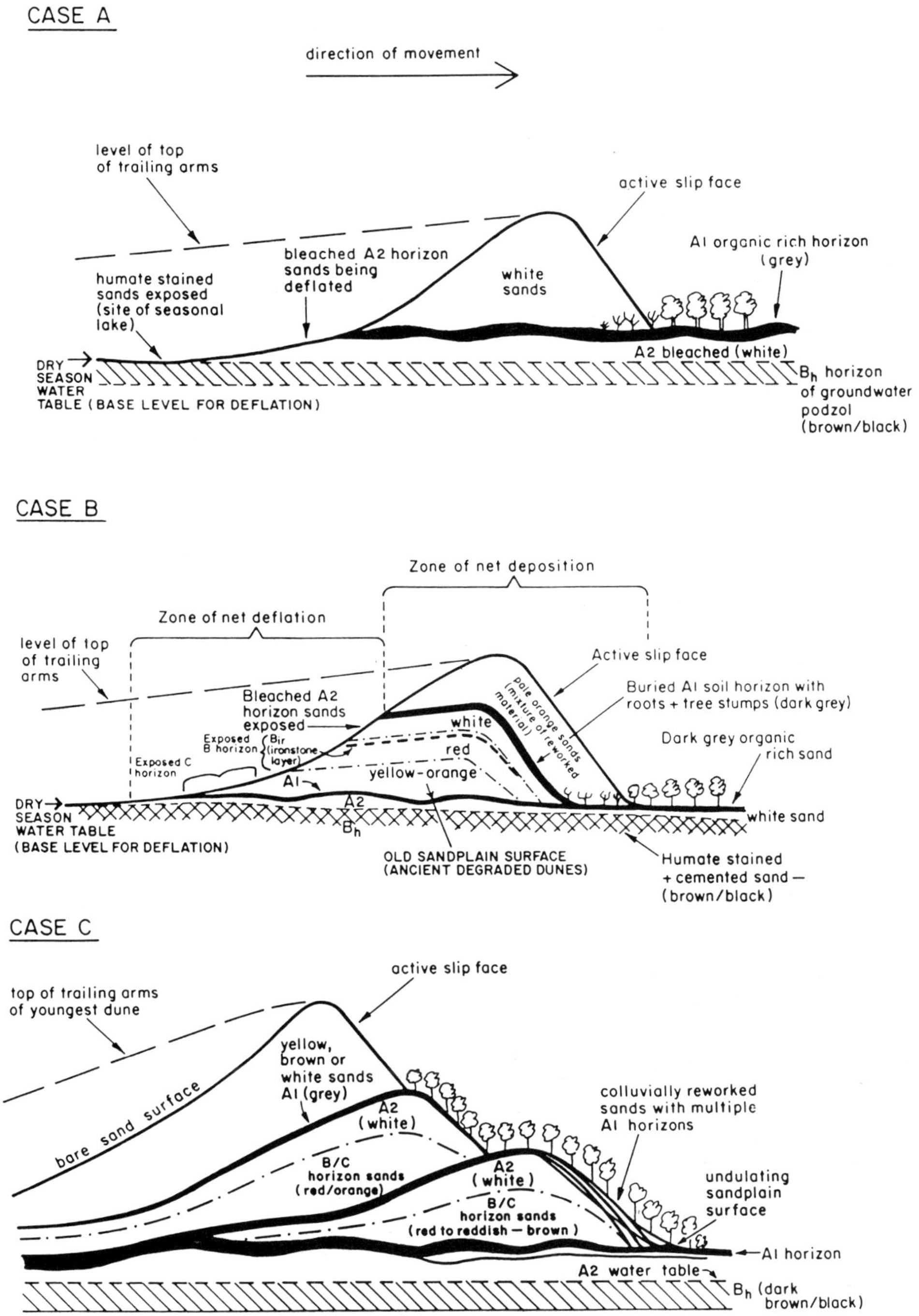

Fig. 14. Schematic diagram showing three contrasting dune stratigraphies in the Cape Flattery area, based on the results of field drilling.

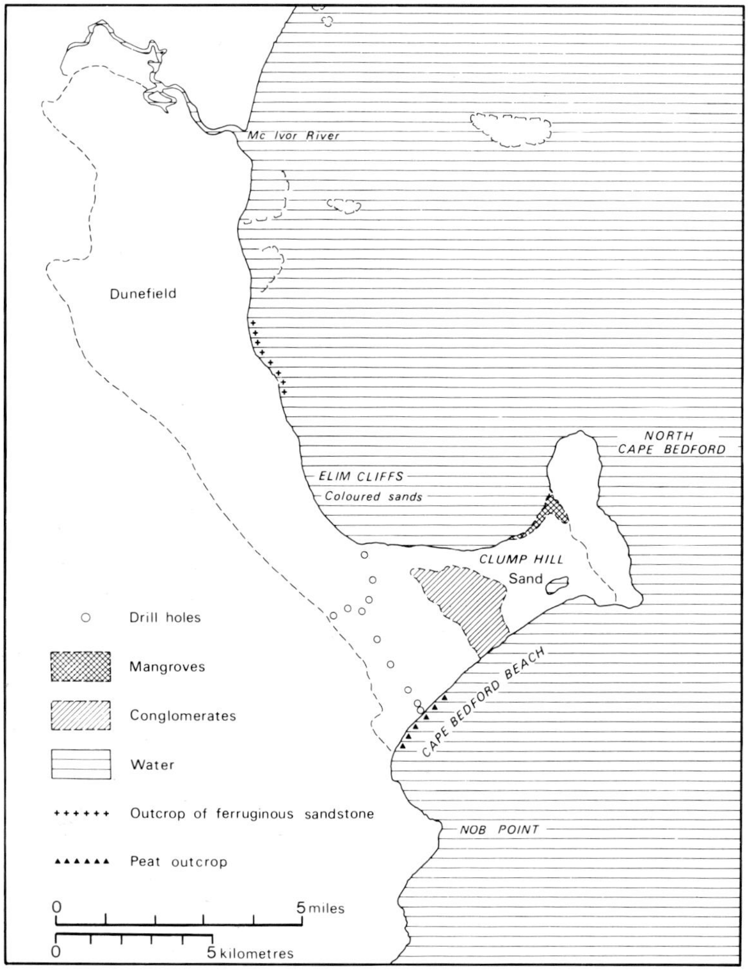

Fig. 15. Map showing coastal sandrock and peat outcrops in the Cape Bedford dunefield.

On the southern side of Shelburne Bay at Double Point, a thin composite layer of aeolian sand (2–5 m thick) overlies bedrock. Lees *et al.* (1990) recognized an upper unit of white sand, interpreted as modern, and a lower unit of brown, humic-rich sand. The latter gave a coarse-fraction (90–125 μm) TL age of 23 800 ± 2400 BP (Xian Au-06).

The southeastern part of the Temple Bay dunefield comprises a series of beach ridge and back-barrier lagoon deposits which have been partially buried and reworked by a transgressive dune sheet (Pye, 1983b). Shell hash from the youngest beach ridge underlying this transgressive dune sheet gave a reservoir-corrected ^{14}C age of 930 ± 135 yrs BP (β-5045). This date provides a maximum age for the overlying dunes, and is consistent with Pye's (1983d) conclusion, based on air photograph measurements and assumed rates of dune movement, that aeolian reworking of these deposits may have begun 300–500 yrs ago.

Cobourg Peninsula dunefield

Samples collected for dating from stabilized dunes to the south of De Courcy Head on the Cobourg Peninsula, Arnhem Land, Northern Territory, indicated the presence of four Holocene dune units (designated A–D) which overlie more indurated sands presumed to be of Pleistocene age (Unit E) (Lees *et al.*, 1990). The currently active dunes (Unit A) were not sampled for dating. The oldest non-indurated dune unit (Unit D) yielded a TL age of 8600 ± 1400 BP (Au-13). The partially superimposed Unit C yielded TL ages of 2600 ± 300 BP (Au-17)

Fig. 16. An 80 m high coastal cliff section at Elim, northwest of Cape Bedford. The dune sand sequence contains at least four major erosional disconformities (bounding surfaces) which correspond to the passage of transgressive dunes. The colour of the sands is highly variable, ranging from white to brown and dark red.

and 2800 ± 500 BP (Au-16), while Unit B gave an age of 1900 ± 400 BP (Au-15). A sample of sand collected immediately beneath a residual pisolitic lag deposit on the surface of a degraded older dune (interpreted as Unit E), located to the south of the dated Unit D dune, gave a TL age estimate of 9300 ± 1100 BP, but this is almost certainly a gross underestimate due to pedogenic concentration of U and Th in and adjacent to the pisolitic layer (Lees *et al.*, 1990). A second sample from greater depth below a similar pisolitic layer in comparable sandy deposits south of Brogden Point, on the southern Cobourg Peninsula, gave a TL age estimate of 81 400 ± 8500 BP (Au-12). This may also be an underestimate of the true age.

Other Queensland dunefields

At Ramsay Bay, on Hinchinbrook Island, dating of backbarrier mangrove peats associated with the dunes indicated that the dune barrier was in existence at least by 9000 ^{14}C yrs BP. Two major phases of parabolic dune activity were identified, one associated with the Holocene marine transgression

Fig. 17. A recent parabolic dune formed by blowout development in a vegetated sandplain surface, 4 km inland from Fourteen Mile Beach, west of Cape Flattery. The vegetation at this site is composed mainly of acidophyllous shrub species, 2–3 m high (Pye & Jackes, 1981).

(9500–6000 ^{14}C yrs BP) and one relating to an apparent input of new sand to the nearshore and dune system after about 900 ^{14}C yrs BP (Pye, 1982b; Pye & Rhodes, 1985). Radiometric evidence from a number of embayments in northern New South Wales also suggests that onshore migration of climbing dunes was associated with the early Holocene marine transgression (Pye & Bowman, 1984).

In southern Queensland, eight dune morphostratigraphic units and eight beach sand units have been identified based on air photograph interpretation and auger drilling to determine the depth of podzolization (Ward & Little, 1975; Ward, 1977). Tentative age estimates were made based on a correlation with marine shorelines in Victoria, but no direct age determinations were presented. Grimes (1979) reported a ^{14}C date of >39 300 yrs BP (SUA-860) for a tree trunk embedded in Ward's Poyungan beach sand at Triangle Cliff, located in the northwestern part of Fraser Island. Two dates of 5690 ± 120 ^{14}C yrs BP (SUA-859) and 4480 ± 115 yrs BP (SUA-858) were obtained respectively from a wood stump and peat in a peat bed within Ward's 'Woorim' beach sand at the same locality. A sample of charcoal from a sample of soil developed on the Poyungan beach sand, and lying beneath Ward's 'Triangle Cliff' dune sand, which gave a date of 3880 ± 105 ^{14}C yrs BP (SUA-433), was interpreted by Grimes as giving a maximum age for the time when Triangle Cliff dune sand transgressed over the Poyungan beach sand in this area.

Subsequent reinterpretation of the field evidence by Ward & Grimes (1987) led to the conclusion that the peat bed dated by Grimes, which was designated 'Bed T', is not part of the Woorim beach sand unit but is more likely to be a time equivalent of the Wathumba beach sand. Moreover, it was agreed that stratigraphic evidence from eastern Fraser Island indicates that the Triangle Cliff dune unit in this area is older than the Wathumba beach sand unit. Ward & Grimes also reported a new date of 6740 ± 80 ^{14}C yrs BP (B-3301) for carbonaceous plant material contained within stabilized Triangle Cliff slip face sands at Triangle Cliff. The date of 3880 ± 105 ^{14}C yrs BP previously obtained by Grimes from the palaeosol beneath the Triangle Cliff dune sand was reinterpreted as representing intrusive roots of plants which grew on the eroded face of the Triangle Cliff dune sand some time after its stabilization. It was therefore concluded by Ward & Grimes (1987) that this dune unit was initiated at a time of lower Holocene sea level and migrated landwards during the early to mid-Holocene marine transgression. The unit is therefore considered by these authors to be time transgressive, underlying and older than the Wathumba beach sand unit on the east side of Fraser Island but overlying and younger than it on the western side of the island.

The entire basis of the scheme presented by Ward & Grimes has been seriously questioned by Coventry (1988), who argued that insufficient evidence is available to support the recognition of the different morphostratigraphic units. The two principal criteria used by Ward (1977) in his initial mapping work, dune morphology and depth to the base of the podzolic A2 horizon, are in themselves an insufficient basis on which to recognize discrete phases of aeolian sedimentation. Thompson's (1981, 1983) work at Cooloola in southern Queensland has demonstrated that the extent of podzolization, reflected by the depth to the base of the bleached horizon, is strongly dependent on the height of a dune and the position of the water table. The latter is often controlled by the nature of the surrounding topography and the composition of the underlying sediments. Dune size and morphology are also dependent partly on the underlying topography, in particular whether the advancing dunes move over and partially rework older dune sand units. The nature and rate of podzolization may also be strongly affected by the sand composition, with important differences occurring between dunes composed of 'new' material supplied from the beach and dunes composed of sand reworked from dunes which have already been highly weathered and deeply podzolized. To date, no grain size, mineralogical or geochemical evidence has been presented to support the morphostratigraphic subdivisions and cross-regional comparisons made by Ward and his associates.

Pickett *et al.* (1985) recovered thirty-nine species of *in situ* scleractinian corals from approximately modern sea level beneath a large stabilized parabolic dune on the western side of North Stradbroke Island. The corals occur within thin inter-tidal sediments which were buried as the large dune advanced towards the northwest. Three species of coral yielded a mean ^{230}Th/^{234}U age of 105 000 yrs BP, suggesting the occurrence of a sea level stand close to the present level during isotope Stage 5c. Pickett *et al.* proposed that the overlying dune sands are of comparable age; however, the contact between the dune sands and the coral-bearing sediments was not observed, and the possibility of a hiatus between the two cannot be ruled out. The corals therefore provide only a maximum age for the overlying dune.

The corals overlie at least 40 m of white sands, but it is uncertain whether these represent buried, *in situ* aeolian sands or sands which have been redeposited subtidally.

SUMMARY AND CONCLUSIONS

Stratigraphic evidence clearly indicates a long history of coastal dune building in northeastern Australia, but considerable uncertainty still surrounds the ages of the dunes and the extent to which major aeolian episodes in different areas were coincident. Drilling has shown that in several areas dune sands extend more than 40 m below present sea level, indicating that aeolian activity occurred in at least some places during cold stages of the late Quaternary when sea level was up to 110 m lower than at present. During the Holocene marine transgression, and probably also during earlier transgressions, the margins of the sand masses were trimmed by marine erosion. In several areas there is strong radiocarbon evidence that new transgressive dunes were initiated along, or close to the eroding shorelines. These dunes migrated landwards during the period 10 000–6000 BP, climbing over bedrock hills, alluvial plains and older generations of stabilized dunes. During this process the underlying sediments were partially reworked and incorporated into the active dunes, in places exposing the more indurated podzolic B horizons which acted as local deflation bases.

Dune formation has continued throughout the later Holocene, although generally on a smaller scale than during the early Holocene. The present evidence suggests that the timing of these late Holocene events has varied between dunefields, indicating a response to local factors rather than global or regional change in climate or sea level. At the present day, some dunefields are entirely stabilized while others show varying degrees of aeolian activity. In some dunefields active dunes have been initiated in the past few hundred years well inland of the shoreline (Fig. 17); nearshore sediment budget changes, sea level rise and shoreline erosion due to changing storm frequency can therefore be ruled out as controlling factors in these instances. However, the possible significance of increases in windiness or loss of vegetation cover due to drought, waterlogging, disease or burning remains to be established.

In virtually all of the major dunefields, insufficient detailed stratigraphic evidence is available to be certain about the number and correlation of aeolian sedimentary units present. Only the youngest deposits have so far been dated by radiocarbon and TL methods, and too few dates are available to establish the extent to which aeolian events in different areas were synchronous.

The pattern of weathering and pedogenesis on the stabilized dunes is complex, with significant differences even on different parts of the same dune due to variations in local drainage conditions. Stratigraphic correlation based solely on pedogenic features, such as sand colour and depth of bleaching, is therefore potentially unsound.

The unusually large size of the parabolic dunes and dunefields in northeastern Australia reflects the long history of sand accumulation, involving partial reworking by successively younger generations of dunes. There is no firm evidence to support the suggestion that stronger winds were responsible for the construction of the larger Pleistocene and early Holocene dunes, although evidence of increases in trade wind intensity during glacial cold stages has been provided elsewhere. The generally smaller scale of the late Holocene features may be related more closely to a reduction in sediment supply after sea level reached approximately its present position around 6000 years ago, leading to cessation of shoreline erosion and, in some areas, progradation of beach ridge/foredune plains. In northern Queensland at least, a second factor contributing to stabilization of dunes in the mid-Holocene may have been increased rainfall (cf. Kershaw, 1986). The size of dunes which have developed from inland blowouts in the late Holocene is directly related to the size of the area in which vegetation cover has been disrupted by burning or some other factor, and to the degree of local wind exposure (e.g. Bird, 1974).

ACKNOWLEDGEMENTS

The research upon which this paper is based was funded by the NERC, the Royal Society, the Leverhulme Trust and the Nuffield Foundation. University of Reading PRIS Contribution No. 212.

REFERENCES

Belperio, A.P. (1983) Terrigenous sedimentation in the Central Great Barrier Reef: a model for the Burdekin region. *BMR J. Aust. Geol. Geophys.* **8**, 179–190.

Bird, E.C.F. (1965) The formation of coastal dunes in the humid tropics: some evidence from North Queensland. *Aust. J. Sci.* **27**, 258–259.

Bird, E.C.F. (1971a) The continental shelf of eastern Australia and its influence on coastal geomorphology. *Quaternaria* **14**, 275–284.

Bird, E.C.F. (1971b) The origin of beach sediments on the North Queensland coast. *Earth Sci. J.* **5**, 95–104.

Bird, E.C.F. (1974) Dune stability on Fraser Island. *Qld Naturalist* **21**, 15–21.

Bird, E.C.F. & Hopley, D. (1969) Geomorphological features on the North Queensland coast. *Aust. Geog. Stud.* **7**, 89–108.

Chappell, J., Chivas, A., Wallensky, E., Polach, H.A. & Aharon, P. (1983) Holocene palaeoenvironmental changes, Central to North Great Barrier Reef inner zone. *BMR J. Aust. Geol. Geophys.* **8**, 223–235.

Chappell, J. & Grindrod, J. (1984) Chenier plain formation in northern Australia. In: *Coastal Geomorphology in Australia* (Ed. Thom, B.G.) pp. 197–231. Academic Press, Sydney.

Coaldrake, J.E. (1961) The ecosystem of the coastal lowlands ('Wallum') of southern Queensland. *CSIRO Bull. 283*, 119 pp.

Coaldrake, J.E. (1962) The coastal dunes of southern Queensland. *Proc. Roy. Soc. Qld* **72**, 101–116.

Cook, P.G. (1986) A review of coastal dunebuilding in eastern Australia. *Aust. Geog.* **17**, 133–143.

Cooper, W.S. (1958) Coastal sand dunes of Oregon and Washington. *Geol. Soc. Am. Mem.* **72**.

Coventry, R.J. (1988) Discussion: history of coastal dunes at Triangle Cliff, Fraser Island, Queensland. *Aust. J. Earth Sci.* **35**, 579–583.

Davies, J.L. (1974) The coastal sediment compartment. *Aust. Geog. Stud.* **12**, 139–151.

Frankel, E. (1974) Recent sedimentation in the Princess Charlotte Bay area, Great Barrier Reef Province. *Proc. 2nd Int. Coral Reef Symp. 2, Brisbane*, 355–369.

Gentilli, J. (1971) The main climatological elements. In: *World Survey of Climatology 13* (Ed. Gentilli, J.) pp. 119–188. Elsevier, Amsterdam.

Grimes, K.G. (1979) Carbon-14 dates and the evolution of Fraser Island. *Qld Govt Min. J.* **80**, 79–82.

Kershaw, P. (1986) Climatic change and Aboriginal burning in north-east Australia during the last two glacial/interglacial cycles. *Nature* **322**, 47–49.

Lavarack, P.S. & Stanton, J.P. (1977) Vegetation of the Jardine River catchment and adjacent coastal areas. *Proc. Roy. Soc. Qld* **88**, 39–48.

LAYCOCK, J.W. (1978) North Stradbroke Island. *Univ. Qld Dept Geol. Pap.* **8**, 89–96.

LEES, B.G., LU, Y. & HEAD, J. (1990) Reconnaissance thermoluminescence dating of northern Australian coastal dune systems. *Quat. Res.* **34**, 169–185.

LITTLE, I.P., ARMITAGE, T.M. & GILKES, R.J. (1978) Weathering of quartz dune sands under subtropical conditions. *Geoderma* **20**, 225–237.

LOURENSZ, R.S. (1977) *Tropical Cyclones in the Australian Region. July 1909–June 1975*. Meteorological Summary for the Bureau of Meteorology. Australian Government Publishing Service, Melbourne, 111 pp.

MAXWELL, W.G.H. (1968) *Atlas of the Great Barrier Reef*. Elsevier, Amsterdam, 258 pp.

MCKEE, E.D. (1979) Introduction to a study of global sand seas. In: *A Global Study of Sand Seas* (Ed. McKee, E.D.) pp. 1–19. US Geol. Surv. Prof. Pap. 1052.

PICKETT, J.W., THOMPSON, C.H., KELLEY, R.A. & ROMAN, D. (1985) Evidence of high sea level during isotope stage 5c in Queensland, Australia. *Quat. Res.* **24**, 103–124.

PYE, K. (1981) Rate of dune reddening in a humid tropical climate. *Nature* **290**, 582–584.

PYE, K. (1982a) Morphological development of coastal dunes in a humid tropical environment, Cape Bedford and Cape Flattery, North Queensland. *Geog. Ann.* **64A**, 213–227.

PYE, K. (1982b) Morphology and sediments of the Ramsay Bay sand dunes, Hinchinbrook Island, North Queensland. *Proc. Roy. Soc. Qld* **93**, 31–47.

PYE, K. (1982c) Characteristics and significance of some humate-cemented sands (humicretes) at Cape Flattery, Queensland, Australia. *Geol. Mag.* **119**, 229–236.

PYE, K. (1983a) Formation and history of Queensland coastal dunes. *Z. Geomorph. Supp. Bd.* **45**, 175–204.

PYE, K. (1983b) The coastal dune formations of northern Cape York Peninsula, Queensland. *Proc. Roy. Soc. Qld* **94**, 33–39.

PYE, K. (1983c) Dune formation on the humid tropical sector of the North Queensland coast. *Earth Surf. Proc. Landf.* **8**, 371–381.

PYE, K. (1983d) Post-depositional reddening of late Quaternary coastal dune sands, northeastern Australia. In: *Residual Deposits* (Ed. Wilson, R.C.L.) pp. 117–129. Geol. Soc. Lond. Spec. Publ. 35. Blackwell Scientific Publications, Oxford.

PYE, K. (1983e) Formation of quartz silt during humid tropical weathering of dune sands. *Sed. Geol.* **34**, 267–282.

PYE, K. (1984) Models of transgressive dune building episodes and their relationship to Quaternary sea level changes: a discussion with reference to evidence from northeastern Australia. In: *Coastal Research: UK Perspectives* (Ed. Clark, M.) pp. 81–104. Geo Books, Norwich.

PYE, K. (1985) Controls on fluid threshold velocity, rates of aeolian sand transport and dune grain size parameters along the Queensland coast. In: *Proceedings of the International Workshop on the Physics of Blown Sand, Aarhus, 28–31 May 1985, Volume 3* (Ed. Barndorff-Nielsen, O.E., Moller, J.T., Romer Rasmussen, K. & Willetts, B.B.) pp. 483–509. Department of Theoretical Statistics, Institute of Mathematics, University of Aarhus, Memoirs 8.

PYE, K. & BOWMAN, G.M. (1984) The Holocene marine transgression as a forcing function in episodic dune activity on the eastern Australia coast. In: *Coastal Geomorphology in Australia* (Ed. Thom, B.G.) pp. 179–196. Academic Press, Sydney.

PYE, K. & JACKES, B. (1981) Vegetation of the coastal dunes at Cape Bedford and Cape Flattery, North Queensland. *Proc. Roy. Soc. Qld* **92**, 37–42.

PYE, K. & RHODES, E.G. (1985) Holocene development of an episodic transgressive dune barrier, Ramsay Bay, North Queensland, Australia. *Mar. Geol.* **64**, 189–202.

PYE, K. & SWITSUR, V.R. (1981) Radiocarbon dates from the Cape Bedford and Cape Flattery dunefields, North Queensland. *Search* **12**, 225–226.

PYE, K. & TSOAR, H. (1990) *Aeolian Sand and Sand Dunes*. Unwin Hyman, London, 396 pp.

RHODES, E.G. (1982) Depositional model for a chenier plain, Gulf of Carpentaria, Australia. *Sedimentology* **29**, 201–221.

ROY, P.S. & THOM, B.G. (1981) Late Quaternary marine deposition in New South Wales and southern Queensland — an evolutionary model. *J. Geol. Soc. Aust.* **28**, 471–489.

SMART, J., POWELL, B.S. & POWELL, D.L. (1974) Auger drilling, northern Cape York Peninsula, Queensland. *BMR J. Aust. Geol. Geophys.* **1**, 211–218.

SWAN, B. (1979) The presence of sand dunes in a tropical low energy zone, Torres Strait (Australia). *Rev. Geomorph. Dyn.* **28**, 61–72.

SWAN, B. (1981) Dunes of Friday Island, Torres Strait, North Queensland. *Singapore J. Tropical Geog.* **2**, 114–128.

THOM, B.G., BOWMAN, G.M. & POLACH, H. (1978) *Holocene Age Structure of Coastal Sand Barriers in New South Wales, Australia*. Dept of Geography, Faculty of Military Studies, University of New South Wales, Duntroon, 86 pp.

THOMPSON, C.H. (1981) Podzol chronosequences on coastal dunes of eastern Australia. *Nature* **291**, 59–61.

THOMPSON, C.H. (1983) Development and weathering of large parabolic dune systems along the subtropical coast of eastern Australia. *Z. Geomorph. Supp. Bd.* **45**, 205–225.

THOMPSON, C.H. & BOWMAN, G.M. (1984) Subaerial denudation and weathering of vegetated coastal dunes in eastern Australia. In: *Coastal Geomorphology in Australia* (Ed. Thom, B.G.) pp. 263–290. Academic Press, Sydney.

THOMPSON, C.H. & HUBBLE, G.D. (1977) Sub-tropical podzols (spodosols and related soils) of coastal eastern Australia. *Proc. Climatrops Conf., Kuala Lumpur*, 203–213.

WARD, W.T. (1977) Sand movement on Fraser Island: a response to changing climates. *Univ. Qld Anthropology Museum Occ. Pap.* **8**, 113–126.

WARD, W.T. (1982) Formation and development of the shoreline at Tangalooma Point, Moreton Island. *Proc. Roy. Soc. Qld* **93**, 11–20.

WARD, W.T. & GRIMES, K.G. (1987) History of coastal dunes at Triangle Cliff, Fraser Island, Queensland. *Aust. J. Earth Sci.* **34**, 325–333.

WARD, W.T. & LITTLE, I.P. (1975) Times of coastal sand accumulation in southeast Queensland. *Proc. Ecol. Soc.*

Qld **9**, 313–317.
WARD, W.T., LITTLE, I.P. & THOMPSON, C.H. (1979) Stratigraphy of two sandrocks at Rainbow Beach, Queensland, and some notes on their composition. *Palaeogeog., Palaeoclimatol., Palaeoecol.* **26**, 305–316.
WILLMOTT, W.F. & POWELL, B.S. (1977) *Explanatory Notes: 1:250 000 Geological Sheet, Cape Weymouth, Queensland*. Bureau of Mineral Resources, Canberra.
WILSON, I.G. (1972) Universal discontinuities in bedforms produced by the wind. *J. Sed. Petrol.* **42**, 667–679.
WRIGHT, A.H. (1976) Genesis of low iron bauxite, north-eastern Cape York, Queensland, Australia. *Econ. Geol.* **71**, 1526–1532.

Spec. Publs Int. Ass. Sediment. (1993) **16**, 45–60

The modern and ancient pattern of sandflow through the southern Namib deflation basin

I. CORBETT

Department of Geology, University of Cape Town, Rondebosch 7700, South Africa

ABSTRACT

The aeolian system within the 0–20 km wide coastal tract of the southern Namib deflation basin is governed by unimodal southwesterly to south-southeasterly surface windflow. A strong seasonality characterizes the wind energy of the system, with the peak occurring from September to March, when gusting at 80–100 km h^{-1} is common.

Despite the high energy of the aeolian system, beach deflation supplying sediment to the aeolian system is restricted to log-spiral and south-facing re-entrant embayments, along the otherwise exposed Atlantic coastline. Sand trap data from a traverse perpendicular to sandflow some 30 km downwind of one such point source prove that rather than fanning out, the sandflow remains within a 1–2 km wide linear, wind-parallel zone. This zone and other similar linear zones are delineated by barchan dune trains which extend up to 120 km from the coast to the Namib Sand Sea. The high-sandflow zones thus act as aeolian transport corridors, in which optimal conditions for aeolian creep transport occur. The possible existence of longitudinal vortices within the planetary boundary layer above the deflation basin provides one mechanism by which the corridors might be maintained.

The distribution of yardangs formed from Precambrian dolomite on the floor of the present-day deflation basin supports the conclusion that the position of aeolian transport corridors has varied with time. Palaeoyardang surfaces, lying outside the present aeolian transport corridors, are undergoing modification by solution weathering. These features must, however, have previously lain within the path of an aeolian transport corridor generated from a regressive coastline. Thus the distribution of aeolian transport corridors has been, and still is, governed by coastal morphology, which is a function of sea level fluctuation and deflation basin morphology. Changes of this sort provide a key to understanding the periodic expansion and contraction of the post-Miocene sand sea, as evidenced by the ongoing formation of erosional, regional bounding surfaces.

INTRODUCTION

The Namib is a coastal desert located on the western margin of the southern African subcontinent. It is bounded to the west by the Atlantic Ocean and to the east by the Great Escarpment (Fig. 1). Although the dynamics of the dune systems comprising the main Namib Sand Sea have been the subject of many studies (e.g. Fryberger & Dean, 1979; Ward, 1984; Lancaster, 1985), the deflation basin, which lies within a restricted diamond area, has not received the same amount of attention. Studies of factors controlling dune morphology and the development and maintenance of the Namib Sand Sea have therefore not fully appreciated the role of the deflation basin, which is an integral part of the aeolian system as a whole.

Prior to this study, the sediment budget has been examined using estimates from wind data of the potential sandflow through the deflation basin (e.g. Fryberger & Dean, 1979; Lancaster, 1985). Such an approach fails to provide a clear understanding of the link between deflation basin dynamics and the aeolian system as a whole because cognizance of factors such as sediment supply which impose

Present address: De Beers Marine (Pty) Ltd, PO Box 2605, Cape Town 8000, South Africa.

constraints on an aeolian transport system no matter how high-energy it may be, cannot be accounted for. This paper is based on 4 years of sandflow data collected along an east–west traverse perpendicular to the dominant sandflow direction through the southern Namib deflation basin. Four sand traps were deployed together with one anemometer station (Fig. 1C). The sand traps were aligned to catch the northbound sandflow only, and their design was based on that of Illenberger & Rust (1986) and Horikawa & Shen (1960). A modified version of the Horikawa & Shen design was compartmentalized to examine the vertical variation of sandflow past a 1.83 m column. The results from this trap were used to calibrate those of the other traps, so that all sandflow figures represent the amount of material transported up to 1.83 m above the bed. The results illustrate the spatial and temporal variations in the distribution of sandflow through the present-day deflation basin, where one of the world's most dynamic aeolian systems currently operates (Rogers, 1977).

Geomorphological evidence shows that changes have occurred in the sandflow pattern through the deflation basin in the past, which can be interpreted on the basis of the model explaining the present-day sandflow distribution. This approach has identified a relationship between sand sea accumulation and eustatic sea level movement, which has implications for studies seeking to relate the dynamics of large aeolian sand bodies to the concepts of seismic sequence stratigraphy.

MORPHOLOGY OF THE SOUTHERN NAMIB DEFLATION BASIN

The deflation basin is wedge-shaped in plan, with its apex in the south situated on the south side of

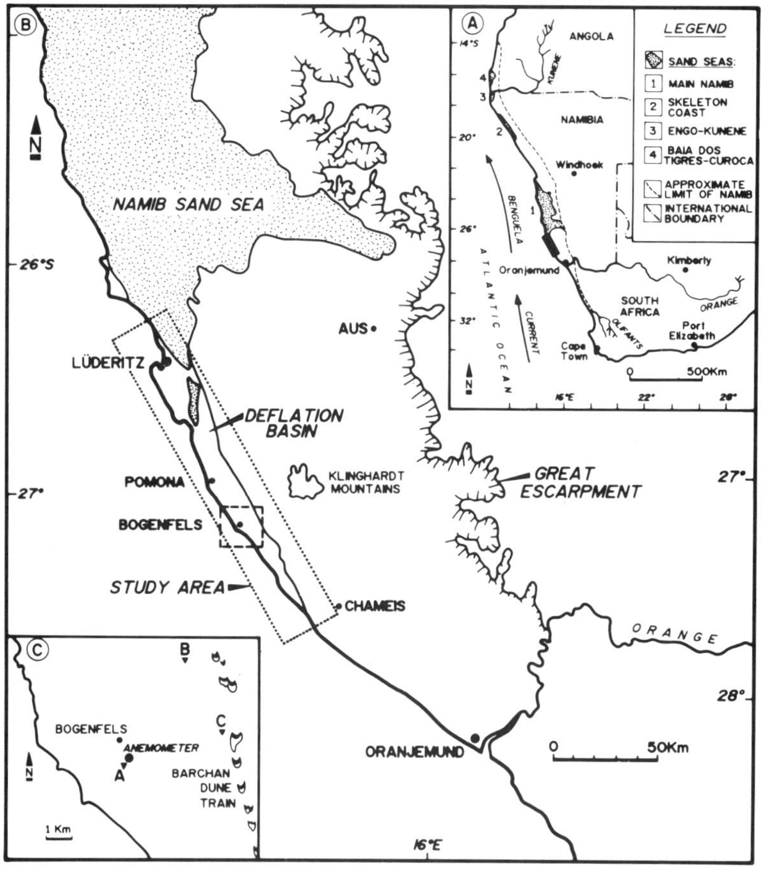

Fig. 1. (A)–(C) Maps showing the location of the southern Namib deflation basin and the position of the anemometer and (▼) sand traps used in this study.

Chameis Bay. It is about 20 km wide to the east of Lüderitz, about 125 km to the north, where it meets the Namib Sand Sea.

In contrast to the Namib Sand Sea, minimal loose sand covers the deflation basin floor, and the regional north–south structural trend which has greatly influenced the development of the rugged deflation basin topography is clearly seen (Fig. 2). Many of the south–north oriented sub-basins within the main deflation basin are endoreic, with no direct connection with the sea (Fig. 3). The largest endoreic basins can be traced over 35 km from south to north, and salt pans commonly occupy base-level locations within them (see Fig. 2). Many of the sub-basins are floored by, and developed within, clastic lithologies of the Bogenfels Formation which are susceptible to mechanical breakdown by salt weathering; the products are rapidly removed by aeolian abrasion and deflation. Consequently, some base-levels have been lowered below the present sea level. The orientation of the endoreic basins thus follows the regional structural trend, which is coincidentally aligned with that of the prevailing southerly surface winds, and the features do not represent yardangs attributable to aeolian processes alone.

Fig. 2. Oblique aerial view looking south over the deflation basin. The strong structural trend of the Late Precambrian Bogenfels Formation is clearly visible.

Fig. 3. Oblique aerial view to the north showing the Idatal, an endoreic basin aligned approximately parallel to the prevailing southerly surface winds.

PATTERN OF SURFACE WINDFLOW

The pattern of surface windflow over the southern Namib can be subdivided spatially into three coast-parallel zones on the basis of changes in the distribution and orientation of aeolian bedforms.

Coastal high-energy regime

The narrow unimodal regime of the 20 km wide coastal tract containing the deflation basin is dominated by high-energy south-southwesterly to south-southeasterly surface winds throughout most of the year (Fig. 4). Occasional reversals, which characteristically result in high-energy northerly winds for brief periods between April and August, are, however, an important seasonal component.

The southerly component is governed by the South Atlantic anticyclonic system, which displays strong seasonal and diurnal variations in response to changes in the pressure gradient which result from the differential heating and cooling of the sea and land surfaces. The peak energy of this component, which is primarily responsible for driving the aeolian system, occurs seasonally between October and March each year. The diurnal peak, when gusting at 80–100 km h^{-1} is most common, occurs between 13:00 and 16:00 hours. There is also some evidence for a regional south–north velocity gradient in addition to the better defined west–east one over the deflation basin (Rogers, 1977; Lancaster, 1985). Although the former is presently ill defined, the experience of living in the region supports the concept of a wind energy increase towards Pomona before it declines gradually towards the start of the sand sea, where depositional processes dominate.

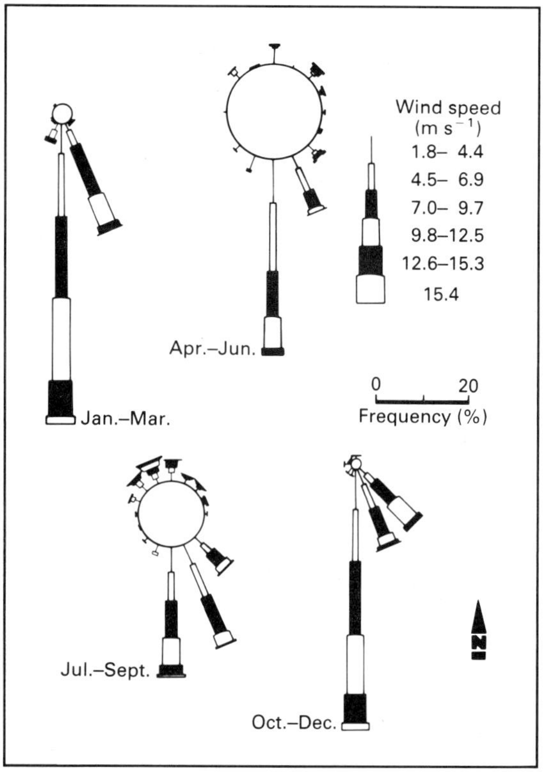

Fig. 4. Wind roses summarizing the seasonal variation in surface wind direction at Bogenfels during 1987, when detailed sand trap results were obtained.

Inland regimes

Two further coast-parallel belts characterized by increasingly complex surface wind regimes occur to the east of the coastal high-energy tract. In both cases, the increasing complexity, resulting in first bimodal and then trimodal regimes, can be attributed to the diminishing influence of the South Atlantic anticyclone towards the interior. In terms of this study, the main effects of these changes are that the wind energy declines in conjunction with the loss of the unimodal nature of the regime away from the coast.

Hence the regional surface wind pattern in the southern Namib broadly resembles that described over the Namib Sand Sea within the central Namib, where the coastal unimodal regime also gives way to bimodal and trimodal regimes to the east, away from the coast (Fryberger & Dean, 1979; Lancaster, 1983, 1985; Ward, 1984).

PRESENT-DAY PATTERN OF SANDFLOW

Coastal sediment entry points and barchan dune trains

The high-energy Atlantic coastline bordering the southern Namib deflation basin is characterized by steep beaches, which are predominantly composed of granules and small pebbles, and rocky shores. Sandy beaches containing abundant material suitable for deflation and subsequent incorporation into the aeolian system by the high-energy southerly surface winds are restricted to four widely separated embayments. Two of the embayments are of a log-spiral form, and the others are south-facing re-entrant bays.

Trains of barchan dunes, which are up to 30 m high and migrate at 30–50 m yr^{-1} (Kaiser, 1926; Corbett, 1989), develop directly to the north of the coastal sediment entry points (Fig. 5), and are a major feature of the southern Namib deflation basin (Kaiser, 1926; Hallam, 1964; O'Brien, 1972; Rogers, 1977). In all cases, the path of the dune trains is remarkably straight. Of the three dune trains which are currently maintained at the coast, the longest can be traced from its inception point at Chameis Bay, through the entire 125 km of the deflation basin, to the point at which it enters the main Namib Sand Sea.

Temporal sandflow variation

The seasonal fluctuation in the quantity of sandflow passing Bogenfels is clearly demonstrated by the sand trap results for site C, which represent a continuous period of about 2 years from 2 December 1986 (Fig. 6A). The same pattern is shown by the results of a detailed monitoring period between 2 December 1986 and 28 May 1987, which confirmed that the sandflow diminishes rapidly from February onwards. Shorter-term fluctuations were also seen within these data, and it was observed that peaks in sandflow do not always correspond with those in surface wind velocity (Fig. 6B). This cyclical pattern of high to low sandflow is interpreted as evidence for the strong influence which sediment availability has on the dynamics of the aeolian system. Within the southern Namib, sediment availability is most likely to fluctuate in response to:

1 seasonal beach progradation, which reaches its maximum between December and February, when the wind energy of the aeolian system is greatest; and/or

2 the storage of sand within the aeolian system itself.

Although beach progradation creates a larger surface area over which deflation can operate, thereby introducing more material to the aeolian system, the incidence of infrequent, high-magnitude 'sand storm' events indicates that sand storage is the more important factor. As shown by Bagnold (1954), sand storage occurs principally in the interstices between clasts on stone pavements and within aeolian current shadows of various scales, including shadow dunes.

The sand storm event which occurred between 10 and 12 February 1987, when the southerly to south-southeasterly surface winds accelerated from an average velocity of about 7 km h^{-1} to 70 km h^{-1} over a few minutes, graphically illustrated the ability of pulses in surface wind energy to entrain material from these storage sites. This single event accounted for 14% of the sandflow measured at site C between 2 December 1986 and 5 January 1988. This contradicts the statement made by Lancaster (1985), who, employing estimates of potential sandflow, concluded that infrequent, high-magnitude 'sand storm' events do not play a significant role in the Namib. This difference in results elucidates the difficulties

Fig. 5. Oblique aerial view to the south along the barchan dune train that migrates to the north past Bogenfels. Sand trap site C was close to the western arm of the 30 m high barchan in the foreground.

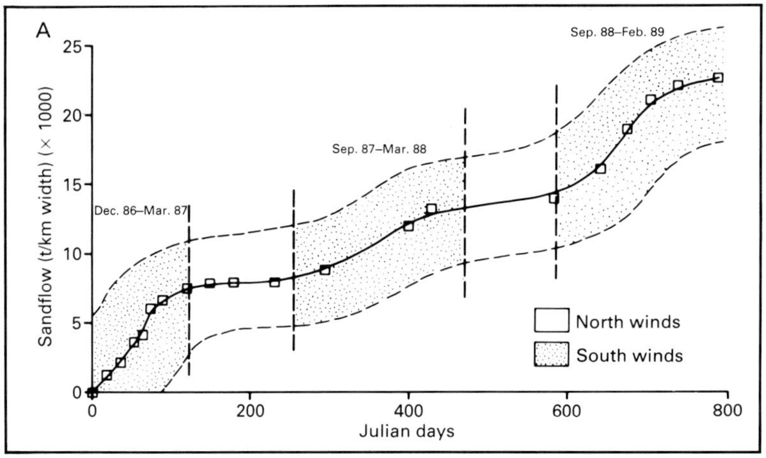

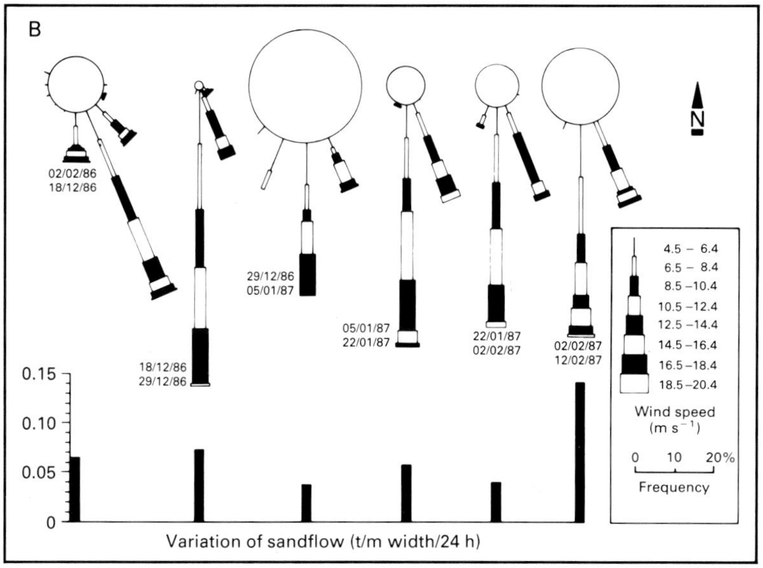

Fig. 6. (A) Graph showing the seasonal variation in the quantity of northbound sandflow (cumulative) transported towards the main Namib Sand Sea, December 1986–February 1989. (B) Detailed sandflow results over short time periods at site C. Cyclical variation in sandflow peaks is seen not to correspond with periods of particularly high wind energy.

involved in calculating meaningful sediment budgets using wind data alone, and indicates a need for more empirically based studies of deflation basin dynamics if aeolian systems are to be more fully understood.

Spatial sandflow variation

The marked spatial variation in the northbound sandflow passing through the deflation basin to the east of Bogenfels is clearly defined by the sand trap results (Fig. 7).

Approximately 7.1 tonnes per metre width (tmw) of sandflow passed site C from 22 January 1987 to 28 May 1987, whilst 0.63 tmw and 0.57 tmw passed sites A and B respectively. A linear zone characterized by high sandflow is therefore associated with the barchan dune train.

The results for site B prove that the western boundary of the high sandflow zone is less than 1200 m from the dune train, and the eastern margin is inferred to lie a similar distance away. The overall width of the zone is therefore about 1–2 km to the east of Bogenfels, some 25 km north of Bakers Bay,

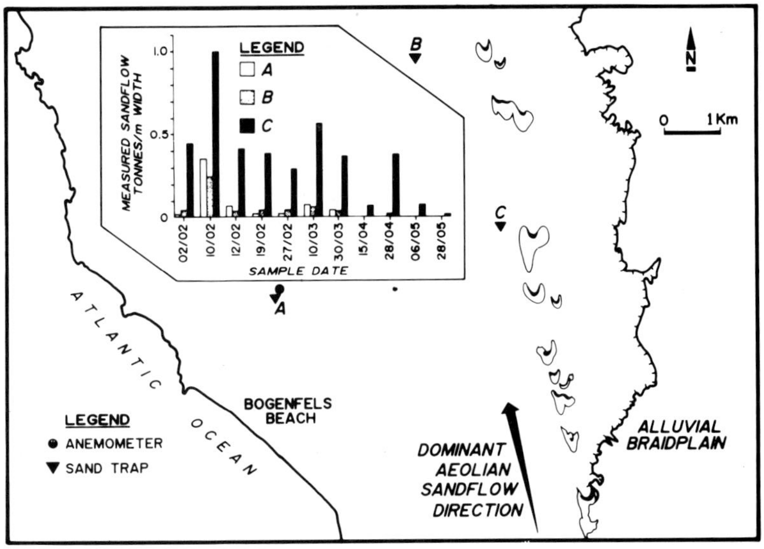

Fig. 7. Graph summarizing spatial variation in sandflow along an east–west traverse across the prevailing southerly surface windflow near Bogenfels.

where numerous workers (e.g. Kaiser, 1926; Hallam, 1964; O'Brien, 1972; Rogers, 1977) have previously shown that the material enters the aeolian system.

AEOLIAN TRANSPORT CORRIDORS

Generation

The length of beach within Bakers Bay along which deflation can act, coupled with the distribution of vegetation, which is restricted to areas of lesser sandflow, show that the width of the high-sandflow zone delineated at Bogenfels is only about 1 km at the coast (Fig. 8). Hence material entering the aeolian system via the deflation of a coastal point source does not fan out downwind to produce an even distribution of sandflow across the entire basin, but remains within a narrow zone, termed an 'aeolian transport corridor', which runs approximately parallel to the southerly surface winds.

Maintenance

Although topographic interaction with the surface windflow undoubtedly influences the path taken by sandflow in some instances, it did not affect the sandflow in the vicinity of the sand trap sites on a subhorizontal alluvial plain beneath a small scarp around 5 m in height. The mechanism whereby aeolian transport corridors exist thus appears to be independent of local topography, and the generation of secondary flow patterns is one process by which this might be achieved.

The arrival of southerly surface winds at the warm land surface of the deflation basin after flowing over the cold surface water of the Atlantic Ocean provides ideal circumstances for the production of considerable thermal contrast. Such conditions are conducive to the development of thermal instability within the planetary boundary layer and, provided that sufficient shear occurs, this can result in the generation of longitudinal spiral vortices (e.g. Jeffreys, 1928; Wilander, 1964). Transverse instability of this type influencing surface windflow is widely acknowledged to be important in the maintenance of linear dune systems (Hanna, 1969; Mabbutt *et al.*, 1969; Glennie, 1970; Folk, 1971), in addition to governing the distribution of dust and pollutants in the atmosphere (Brown, 1983). The occurrence of linear dune systems throughout the southern Namib (Fig. 9) thus supports the concept that secondary flow within the planetary boundary layer is widely developed over the region, with additional evidence being provided by the occasional presence of cloud streets above the deflation basin.

The crestal spacing of the linear dunes to the southeast of the deflation basin varies from 250 to 500 m, whilst in the southern part of the main Namib Sand Sea it is about 2 km, which roughly corresponds to the estimated width of the Bakers Bay aeolian transport corridor. Theoretically, the difference in

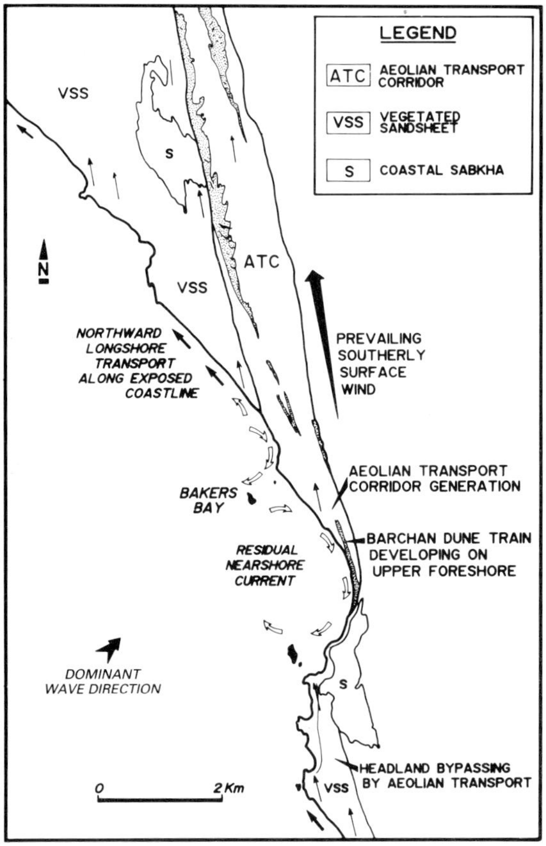

Fig. 8. Sketch showing the Bakers Bay log-spiral embayment, which acts as a point source for sediment entering the Bakers Bay barchan dune train as a result of beach deflation.

the crestal spacing of the different linear dune systems could be due to the compression of the anticyclonic wind system against the Great Escarpment, which forms an orographic barrier. The aeolian transport corridor may thus correspond to the conceptual model of Allen (1982) for secondary flow in a system in which there is insufficient sediment to cover the bed (Fig. 10A). The margins of the aeolian transport corridor on the deflation basin floor would then represent the position along which the longitudinal reattachment lines of adjacent lines of longitudinal vortex cells diverge. Sandflow entering the aeolian system between two longitudinal reattachment lines would then be prevented from fanning out downwind by the secondary flow pattern across the bed, which would direct the material in towards the barchan dune train (Fig. 10B). The maximum areal concentration of sandflow travelling over the bed within the aeolian transport corridor is predicted to occur along the separation line of adjacent longitudinal vortex cells, which is probably delineated by the barchans comprising the dune train. Where sufficient sediment is available a number of parallel barchan dune trains can develop and be maintained within adjacent pairs of longitudinal vortex cells, as seen at Grillental directly to the south of the main Namib Sand Sea (Fig. 11).

Three major aeolian transport corridors currently operate within the southern Namib deflation basin, as evidenced by the development of barchan dune trains. All of these occur downwind of log-spiral embayments at Chameis Bay, Bakers Bay and Prinzenbucht (Fig. 12). A fourth, smaller corridor is generated at the small south-facing re-entrant embayment called Van Reenen Bay. In addition, sandflow is also generated by beach deflation along the coast in the vicinity of Elizabeth Bay, which also provides an important sediment source for the aeolian system. Coastal morphology therefore plays an important role in determining the frequency and spacing of aeolian transport corridors, which are thus susceptible to changes brought about by sea level movement.

During prolonged periods of southerly surface wind, the aeolian dispersal system establishes an equilibrium with the prevailing wind conditions. Provided that abrupt peaks in wind velocity do not occur, there is a progressive build-up in the quantity of saltation load stored within the aeolian system. Brief, high-energy pulses of southerly surface wind rapidly entrain this material, producing infrequent, high-magnitude sand storm events which play an important role in the Namib aeolian system. Sandflow is maintained at much higher levels within aeolian transport corridors during such events, but the influence of aeolian transport corridors on the distribution of sandflow diminishes greatly during infrequent, high-energy, northerly wind reversals. Under these circumstances, abundant sand is rapidly entrained from storage sites which are only stable under southerly wind conditions, with the result that high sandflow occurs basin-wide.

Displacement of aeolian transport corridors

Yardangs are widely distributed throughout the deflation basin, but they are best developed where Late Precambrian Bogenfels Formation dolomite provides upstanding relief due to its resistance to

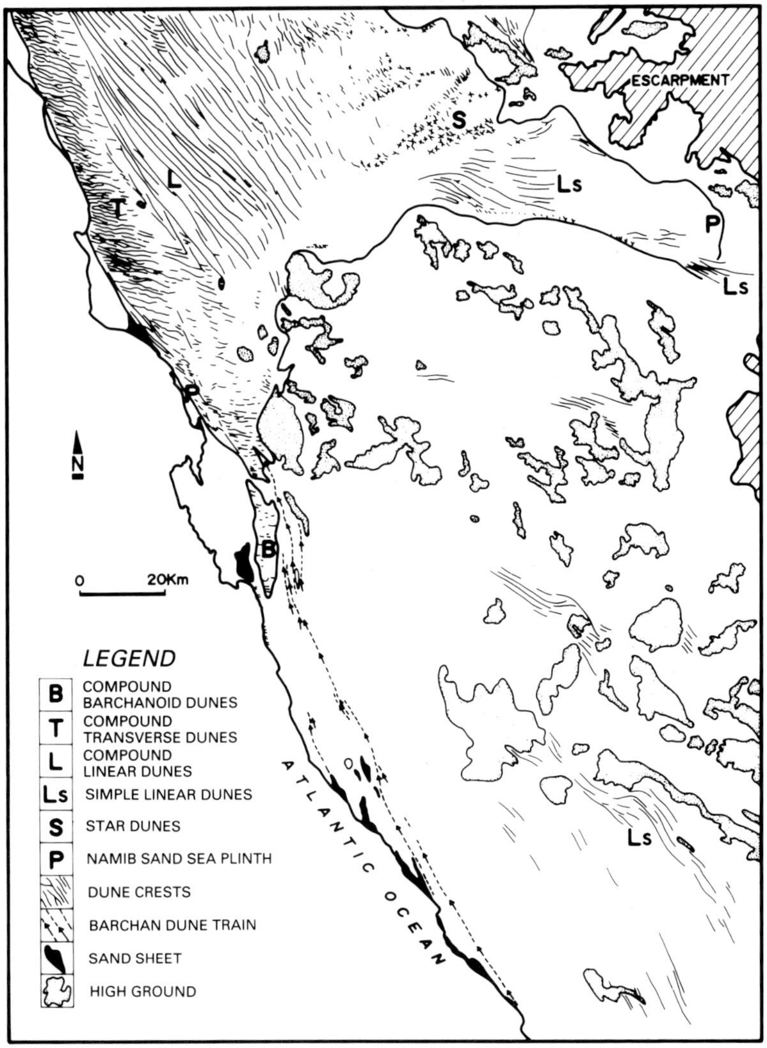

Fig. 9. Map showing the regional distribution of aeolian linear dune systems within the southern Namib.

salt weathering. The yardangs are unusual because they are formed from crystalline rock, and the tectonic structure has had some influence, with the best forms occurring where the dolomite strikes east–west and dips to the north (Fig. 13). Evidence of the extremely abrasive nature of the aeolian system is provided by the development of beautiful ventifacts, with some of the most complex and polished forms being produced from dolomite (Fig. 14).

Areas in which yardang formation is actively occurring today are restricted to the paths of the currently active aeolian transport corridors (Fig. 15). Under these conditions, yardang surfaces are smoothly polished and fluting is very evident. A number of yardang fields do, however, lie beyond the influence of the present-day aeolian transport corridors and they are no longer developing actively. In these cases, the once smoothly polished surfaces are now rough and colonized by lichen, with fluting barely identifiable due to surface modification by solution weathering of the dolomite.

The Bogenfels yardang field is presently situated on the coast, where deflation of the straight, steep, high-energy beach provides minimal sediment input to the aeolian system. This yardang field must therefore have developed during a regression, when bays providing coastal point sources for aeolian transport corridor generation would have lain further

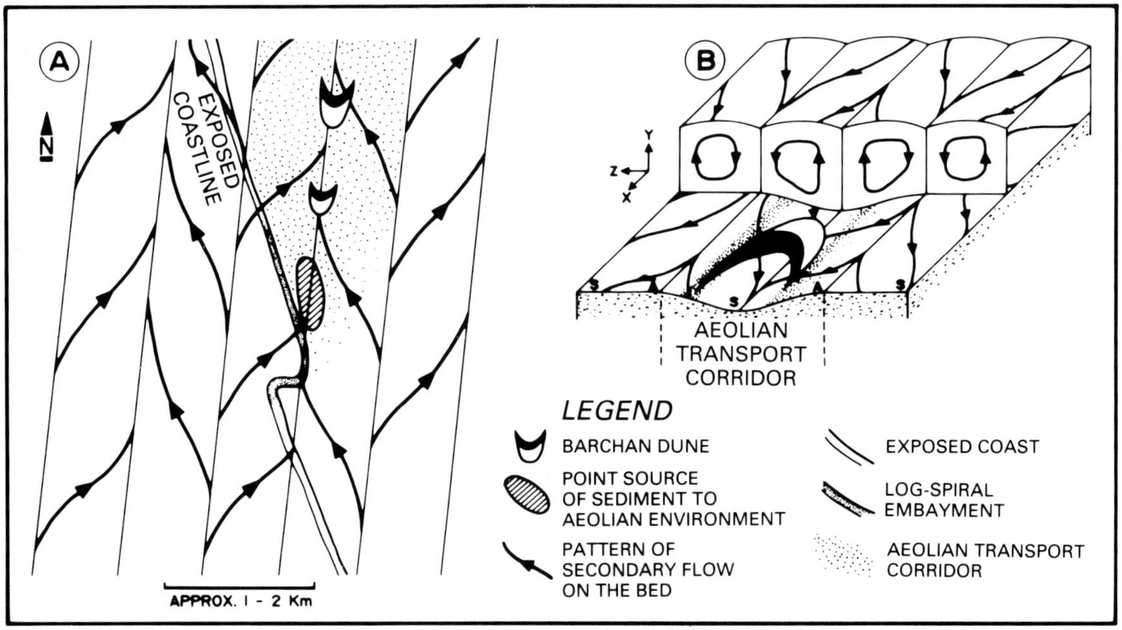

Fig. 10. (A) Conceptualized pattern of helical secondary vortices which are proposed as a mechanism for maintaining aeolian transport corridors. (B) Proposed role of helical secondary flow in constraining sandflow entering an aeolian transport corridor through the deflation of a coastal point source to the pair of vortices which flow across it.

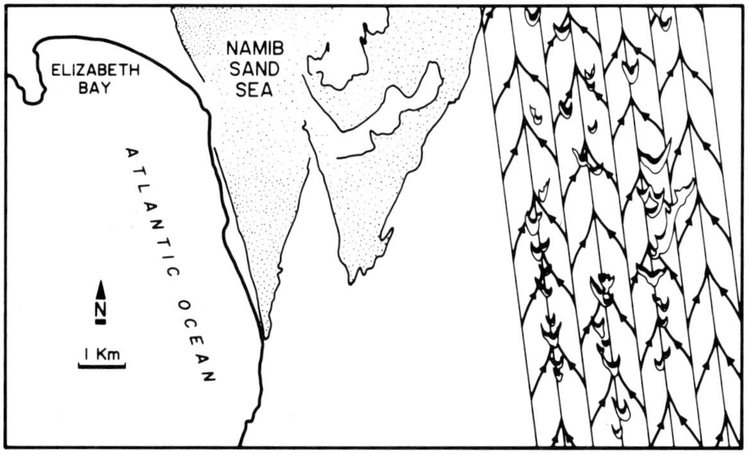

Fig. 11. Adjacent barchan dune trains crossing the Grillental Valley to the south of Kolmanskop. Note the regular spacing which might reflect the presence of helical secondary vortices controlling the spacing of the dunes which define the path of aeolian transport corridors.

west. Conversely, non-active yardangs along the eastern margin of the present-day deflation basin imply a relative shift of the aeolian system to the east. The location of the deflation basin margin is governed by the width of the coastal high-energy tract dominated by southerly surface windflow. In order to account for the margin being displaced to the east, either the width of the coastal tract must be increased or the position of the high-energy tract itself must alter. The latter case might be expected to occur during a transgressive period, when an eastward shift in the relative position of the zones comprising the regional pattern of surface windflow over the Namib might be anticipated.

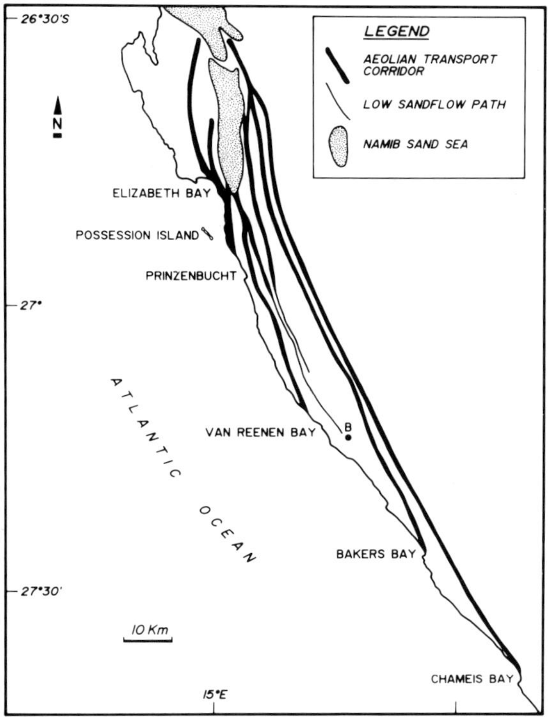

Fig. 12. The present-day distribution of aeolian transport corridors within the southern Namib deflation basin, which are responsible for the maintenance of the main Namib Sand Sea.

Fig. 14. Magnificent example of an aeolian ventifact consisting of blades and spikes of crystalline dolomite. Note the extreme development of the erosional features on the south face of the block. Scale is 10 cm.

Fig. 13. Oblique aerial view to the east showing 10–15 m high yardangs formed from crystalline dolomite immediately west of Bogenfels.

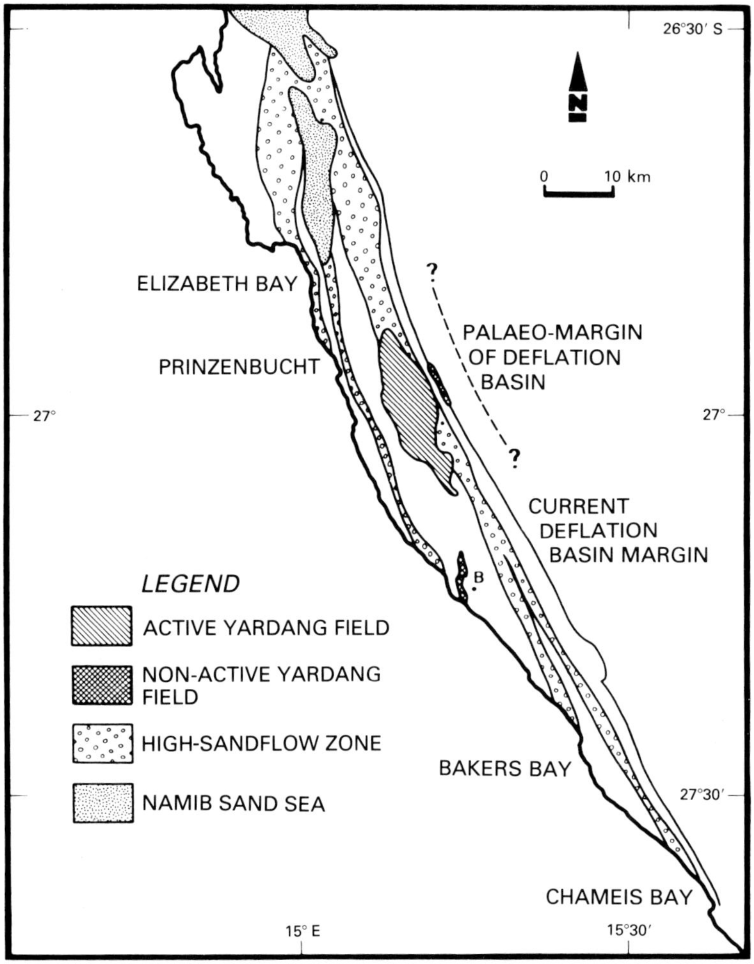

Fig. 15. Map showing the distribution of currently active yardang fields with respect to the present-day pattern of sandflow, together with non-active fields which are no longer maintained by aeolian abrasion.

INFLUENCE OF SEA LEVEL MOVEMENT ON DEFLATION BASIN DYNAMICS AND SAND SEA MAINTENANCE

Extensive regional mapping has shown that at least four palaeodune systems deposited by southerly quadrant palaeowind regimes are represented within the southern and central Namib (Fig. 16) (Ward, 1984, 1987; Corbett, 1989; Ward & Corbett, 1990). In all cases the palaeodune systems were larger than the present-day main Namib Sand Sea (Sossus Sand Formation), and thus provide evidence for sand sea expansion. Although it is difficult to date the palaeodune systems it is believed that sand sea expansion(s) occurred during marine regression(s), when the zones comprising the regional surface windflow pattern may have shifted westwards in response to a change in the position of the shoreline. The highest-energy tract of the present-day system is delineated by the deflation basin and the coastal belt of compound transverse dunes along the western border of the main Namib Sand Sea, which Ward (1984) has shown to be considerably more dynamic than the linear dune systems to the east. Thus, conceptually, a considerably less dynamic eastern zone of sand sea (core) is bordered to the west by increasingly dynamic coast-parallel belts, characterized by a progressively greater throughput of sandflow. These dynamic belts are the most responsive to changes in the aeolian system brought about by changes in deflation basin dynamics related to sea level movement.

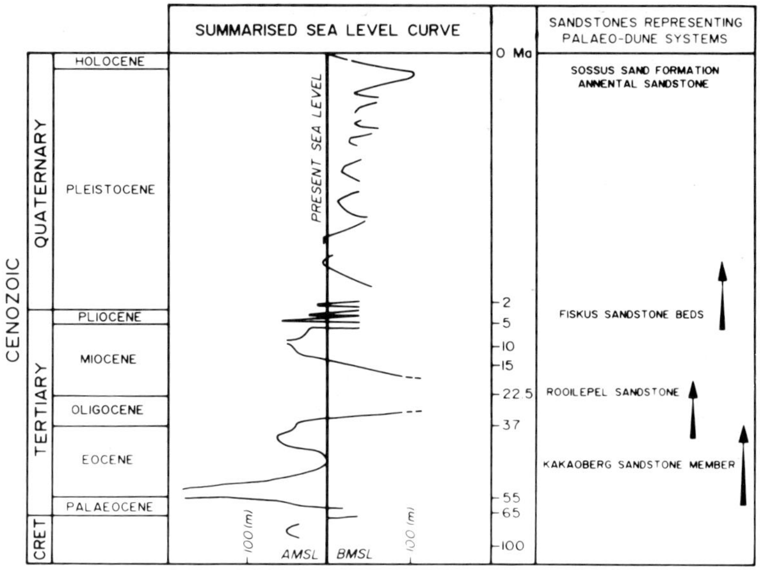

Fig. 16. Lithostratigraphy of palaeodune systems preserved within the southern Namib deflation basin.

Response to regression

During a regression it is believed that the zones of different surface windflow shift to the west, as the continental shelf along the margin of the former sand sea plinth is gradually exposed. Any sandflow arriving at the southern end of the sand sea from the deflation basin is then envisaged to accrete as wedges along the former plinth, with the result that the sand body expands westwards. Although deflation of the newly exposed shelf initially also supplies sediment, it is probably finite due to the development of stable stone pavements and deflation lags. The western belt of high sandflow along the new margin of the sand sea is thus most likely to be maintained by the continued introduction of sandflow via aeolian transport corridors, which may increase or decrease in number depending on changes in coastal morphology. It is predicted that, in time, former belts of compound transverse dunes would be replaced by linear dune forms and, were sections through the palaeodune sequences available, it is suspected that changes of this type would be reflected in the primary sedimentary structure preserved within the sand body.

Response to transgression

Conceptually, the coastal belt of high sandflow migrates conversely to the east during a transgression, with the result that the western margin of the sand sea is reworked not only by nearshore marine processes but by aeolian processes as well. This is probably the current situation of the Namib Sand Sea.

Regional bounding surface formation

Based on present knowledge, two periods of sand sea expansion have occurred since the mid-Miocene. The oldest, represented by the Fiskus Sandstone Beds, was the precursor of the present sand sea, and probably developed from around 10 Ma, whilst the other sand body for which we have some evidence is represented by the Annental sandstone, and was probably deposited during the last glacial maximum, about 18 000–20 000 BP. As a result of the deflation basin environment migrating to the east in response to transgression subsequent to the deposition of these two sand bodies, the present-day aeolian system is actively reworking them at the southern extremity of the main Namib Sand Sea. The Annental sandstone has now been largely eroded, leaving rare isolated remnants in steep sided coastal gulleys and within the deeper ephemeral stream beds, but due to the partial cementation of the Fiskus Sandstone Beds the latter are being eroded more slowly to form an extensive erosional plane. This erosional plane varies from being subhorizontal to forming a gentle incline, and the foresets of former crescentic dune forms are clearly seen on its surface (Fig. 17). It is interpreted as a regional bounding surface, or supersurface, which is in the process of

Fig. 17. Erosional, regional bounding surface being cut in the Fiskus Sandstone Beds as a result of a shift in the location of the southern Namib deflation basin.

developing. The coarse-grained nature of the Fiskus Sandstone Beds is a distinguishing feature of this sand body, which contains substantial quantities of granules within the grainflow strata of cross-bedded sets. The deflation of the sandstone once it has weathered provides material for incorporation into large granule ripples which occur extensively on the regional bounding surface, and the weathering products are also frequently incorporated into ephemeral stream systems which cross the surface.

If the regional bounding surface described above was subsequently covered by sand sea expansion during sea level regression, the absence of datable alluvial or marine deposits separating successive palaeodune sequences would obscure the true nature of the bounding surface. At least one period of sand sea expansion represented by the Annental sandstone would therefore go unrecognized, and evidence for others might also have been removed by aeolian erosion. Under these circumstances the correct incorporation of stacked palaeodune sequences, such as those preserved in the southern Namib Desert, into a sequence stratigraphic framework appears to be difficult to accomplish.

SUMMARY AND CONCLUSIONS

It is concluded that empirical sandflow data are required in order to determine fully the sediment dynamics of deflation basin environments, which are complicated by variations in sediment supply, sediment availability and substrate type. In the case of high-energy systems such as that of the southern Namib deflation basin, calculations of potential sandflow from wind data alone are unable to account for these factors, masking the significance of infrequent, high-magnitude sand storm events for the aeolian system.

Sediment supply to the aeolian system within the southern Namib is restricted to log-spiral and south-facing re-entrant embayments. Material deflated from the beach environment enters the aeolian system via aeolian transport corridors which vary from 1 to 2 km in width, run parallel to the prevailing southerly surface windflow, and are characterized by conditions of very high sandflow. The passage of aeolian transport corridors through the deflation basin is defined by trains of barchan dunes, which migrate along strikingly straight paths. It is suggested that longitudinal secondary vortices within the planetary boundary layer are responsible for the maintenance of aeolian transport corridors, which appear to operate independently of deflation basin topography.

Four aeolian transport corridors currently operate within the deflation basin, with a further high-sandflow zone occurring in the vicinity of Elizabeth Bay, which is probably too close to the Namib Sand Sea to allow full development of an aeolian transport corridor. On arrival at the Namib Sand Sea, sandflow derived from aeolian transport corridors is believed to remain within a narrow, high-energy coastal tract characterized by compound transverse dunes. Much of the material which enters therefore probably bypasses the less dynamic eastern 'core' which forms the majority of the sand body.

The frequency and spatial distribution of aeolian transport corridors are not constant in either time or space. Such changes are brought about by sea level movement, which modifies coastal morphology and thus alters the sites at which log-spiral and south-facing re-entrant bays form through the drowning of portions of endoreic basins which are the main geomorphological elements of the deflation basin. Lateral shifts in the zonal surface wind pattern are predicted to accompany regression and transgression.

Palaeodune systems preserved within the Namib provide evidence for at least three periods of sand sea expansion during regressive events, which is envisaged to occur by the accretion of wedges of material along the former sand sea plinth as the continental shelf is progressively exposed. The western margin of the expanded sand sea is reworked by both nearshore marine and aeolian processes during transgression, which results in sand sea contraction such as can be seen today. Palaeodune systems are influenced by deflation basin dynamics as the zonal components of the surface wind pattern migrate eastwards, which results in extensive erosion on a regional scale to produce an undulating regional bounding surface. Virtually all evidence of sand sea expansion during the last glacial maximum, about 18 000–20 000 BP, has now been removed. The full significance of the regional bounding surface, if it were subsequently to be covered during sand sea expansion, would be virtually impossible to ascertain, and it is concluded that the incorporation of aeolian sequences, such as that preserved in the Namib, within a sequence stratigraphic framework would be very difficult to achieve.

ACKNOWLEDGEMENTS

It is a pleasure to thank Mr J.B. Hawthorne, Dr C.R. Clement and Mr K.R. Hazell of the Anglo American Corporation of South Africa and CDM (Pty) Ltd for their support and permission to undertake this study. I would particularly like to thank the Security Department of CDM (Pty) Ltd for providing the helicopter flights from which this project has benefited greatly. I would also like to thank the corporation for enabling me to attend the conference and to present this paper, and acknowledge the draughtsmanship of Mr B. Griffin's team at De Beers Marine.

The research recounted in this paper forms part of my Ph.D. thesis, which was undertaken at the Department of Geology, University of Cape Town. I am extremely grateful to my supervisor, Professor W.E.L. Minter, for many helpful discussions during the course of the work.

Finally, and most importantly of all, I would like to thank my wife Ali for contributing so much to the completion of this research out in 'the dunes', and for her constant and unfailing support away from the sand.

REFERENCES

ALLEN, J.R.L. (1982) *Sedimentary Structures: Their Character and Physical Basis*. Elsevier, Amsterdam, 2663 pp.

BAGNOLD, R.A. (1954) *The Physics of Blown Sand and Desert Dunes*, 2nd edn. Methuen, London, 265 pp.

BROWN, R.A. (1983) The flow in the planetary boundary layer. In: *Eolian Sediments and Processes* (Ed. Brookfield, M.E. & Ahlbrandt, T.S.) pp. 291–310. Developments in Sedimentology 38. Elsevier, Amsterdam.

CORBETT, I.B. (1989) *The sedimentology of the diamondiferous deflation deposits within the Sperrgebiet, Namibia*. Unpublished Ph.D. thesis, University of Cape Town, 430 pp.

FOLK, R.L. (1971) Longitudinal dunes of the northwestern edge of the Simpson Desert, Northern Territory, Australia. 1. Geomorphology and grain size relationships. *Sedimentology* **16**, 5–54.

FRYBERGER, S.G. & DEAN, G. (1979) Dune forms and wind regime. In: *A Study of Global Sand Seas* (Ed. McKee, E.D.) pp. 136–169. US Geol. Surv. Prof. Pap. 1052.

GLENNIE, K.W. (1970) *Desert Sedimentary Environments*. Developments in Sedimentology 14. Elsevier, Amsterdam, 222 pp.

HALLAM, C.D. (1964) The geology of the coastal diamond deposits of Southern Africa (1959). In: *The Geology of Some Ore Deposits in Southern Africa* (Ed. Haughton, S.H.) pp. 671–728. Spec. Publ. Geol. Soc. S. Afr., Johannesburg.

HANNA, S.R. (1969) The formation of longitudinal sand dunes by large helical eddies in the atmosphere. *J. Appl. Meteorol.* **8**, 874–883.

HORIKAWA, K. & SHEN, H.W. (1960) Sand movement by wind action — on the characteristics of sand traps. *Beach Erosion Board Tech. Mem. 119*, 51 pp.

ILLENBERGER, W.K. & RUST, I.C. (1986) Venturi-compensated eolian sand trap for field use. *J. Sed. Petrol.* **56**, 541–542.

JEFFREYS, H. (1928) Some cases of instability in fluid motion. *Proc. Roy. Soc. London Ser. A* **118**, 195–208.

KAISER, E. (1926) Die jungen sedimentären neubildung om extrem-aridenklima der Namibwüste. In: *Die Diamantenwüste Südewestafrikas 2* (Ed. Kaiser, E.) pp. 317–380. Dietrich Reimer (Ernst Vohsen), Berlin.

LANCASTER, N. (1983) Linear dunes of the Namib Sand Sea. *Z. Geomorph. Suppl.* **45**, 27–49.

Lancaster, N. (1985) Winds and sand movements in the Namib Sand Sea. *Earth Surf. Proc. Landf.* **10**, 607–619.

Mabbutt, J.A., Wooding, R.A. & Jennings, J.N. (1969) The asymmetry of Australian desert sand ridges. *Australian J. Sci.* **32**, 159–160.

O'Brien, R.F. (1972) The barchans of the Southern Namib: grain size analysis. *Proc. 4th S. Afr. Univ. Geogr. Conf.*, 22–31.

Rogers, J. (1977) Sedimentation on the continental margin of the Orange River and the Namib Desert. *Joint Geol. Surv. S. Afr. & Univ. of Cape Town Mar. Geosci. Unit Rep. 7.*

Ward, J.D. (1984) *Aspects of the Cenozoic geology in the Kuiseb River Valley, Central Namib Desert.* Unpublished Ph.D. thesis, University of Natal, Pietermaritzburg, 310 pp.

Ward, J.D. (1987) The Cenozoic succession in the Kuiseb Valley, central Namib Desert. *Mem. Geol. Surv. Namibia* **9**, 1–124.

Ward, J.D. & Corbett, I.B. (1990) Towards an age for the Namib. In: *Namib Ecology: 25 Years of Namib Research* (Ed. Seely, M.K.) pp. 17–26. Transvaal Mus. Monogr. 7.

Wilander, P. (1964) Convective instability in a two layer fluid heated uniformly from above. *Tellus* **16**, 349–359.

Spec. Publs Int. Ass. Sediment. (1993) **16**, 61–69

Internal structure of an aeolian dune using ground-penetrating radar

C.J. SCHENK, D.L. GAUTIER, G.R. OLHOEFT *and* J.E. LUCIUS
US Geological Survey, MS 971, Box 25046, Denver, CO 80225, USA

ABSTRACT

A ground-penetrating radar survey was made on a large complex aeolian dune along the margin of Great Sand Dunes National Monument, Colorado, to delineate the internal structures formed by dune migration in a complex wind regime. Radar waves were partially reflected from sediment interfaces that had differing densities or moisture contents. In this way bounding surfaces between sets could be interpreted from changes in the attitude of sets of reflectors. The radar reflectors were recorded to depths of 15 m, but the best resolution of bounding surfaces was obtained in the upper 5 m of the dune sand. Bounding surfaces interpreted from reflectors define a main dune set 5–8 m thick, with foresets up to 23 m long. Thicknesses of other wedge-shaped and tabular planar sets range from 0.75 to 1.5 m, averaging 1 m; set lengths range from 6 to 12 m, averaging 8.5 m. Trough-shaped sets range in thickness from 0.5 to 3 m, averaging 1.1 m, and range in width from 5 to 22 m, averaging 10 m. These trough structures may have been caused by the migration of scour pits associated with small superimposed dunes, or may be the result of scour fills formed during reversing winds. Reversing winds also formed numerous subtle bounding surfaces (reactivation surfaces) along the leading edge of the dune as it migrated, defining sets ranging in thickness from 0.5 to 2 m, averaging 1 m, and with foreset lengths ranging from 15 to 23 m, averaging 20 m. This study demonstrates the usefulness of ground-penetrating radar in resolving the internal structures of damp to dry, clay-free, aeolian dune sand. The numerous sets and bounding surfaces resolved with radar indicate a relationship between the complexity of internal structure and the multiple directions of sand-carrying or sand-scouring winds. In a tectonically active basin such as the San Luis Valley, these complex aeolian dune sand bodies have excellent preservation potential.

INTRODUCTION

The internal structures of large modern aeolian dunes are known from only a few studies because of the obvious difficulties in excavating or trenching unconsolidated sand. Trenching is generally limited by the depth to the local water table, but most hand-dug trenches seldom extend beyond about 1 m deep. Large trenches were bulldozed through several dunes at White Sands, New Mexico (McKee, 1966), but the excavations were made possible to a large degree by slight cementation of the gypsum sand. Excavations on this scale have not been carried out for quartzose dune sand. These kinds of studies are necessary if we are to further our understanding of the relationships between aeolian dune type, dune internal structure and complexity of effective sand-carrying winds.

Ground-penetrating radar is a technique whereby a large cross-section (tens to hundreds of metres long, several metres deep) can be made that allows interpretation of the three-dimensional internal geometry of some modern sediments without resorting to excavation. The purpose of this study is to demonstrate the application of ground-penetrating radar in determining the internal structures and geometry of a large aeolian dune. Most previous applications of ground-penetrating radar have involved a search for buried objects (pipes, cables, drums of chemicals, etc.) in hazardous waste, construction sites and archaeological investigations (Imai *et al.*, 1987; Daniels *et al.*, 1988), but radar is beginning to be applied to sedimentological problems (Moorman *et al.*, 1990).

As with high-resolution seismic geophysical studies, ground-penetrating radar is a form of remote sensing; internal structures and bounding surfaces are interpreted from processed data rather than being viewed directly. The main thrust of this study was to try to interpret bounding surfaces between sediment packages from the radar record, as bounding surfaces outline the architectural elements of aeolian deposits. Generally, bounding surfaces can be recognized on the radar transect, depending on the wavelength of the radar antenna. The bounding surfaces were of second and third order (Brookfield, 1977; Kocurek, 1988, 1991), or what have been called 'growth surfaces' (Fryberger, 1991). Loss of resolution with lower-frequency antennas meant that bounding surfaces could not be interpreted with certainty below a depth of 5 m.

RADAR OPERATION

Ground-penetrating radar uses radio-frequency electromagnetic energy to measure contrasts in electrical properties in sediment (Duke, 1990; Olhoeft, 1990a). The energy radiated into the ground by an antenna is reflected at interfaces where electrical properties change. When recorded and processed, the two-way travel times produce a cross-sectional image of variations in electrical properties. Radar antennas are built to radiate certain frequencies. The higher the antenna frequency, the shallower the depth of penetration, but the greater the resolution of internal structure. As lower frequencies are used, fewer reflectors are identified, and bounding surfaces and sediment packages are less defined. However, the depth of penetration increases (Fig. 1). In this study, frequencies of 300, 500 and 900 MHz were used, and the records were stacked to produce a final cross-section. The ground-penetrating radar hardware used in this study is commercially available in the USA, but the software is not commercially available for processing data to the extent illustrated in Fig. 1.

In operation, each antenna is pulled at a constant velocity across the sand surface. Location marks (every 10 m) are used to correct the record for variations in the speed of the antenna movement. The line of transect was surveyed with a laser theodolite to correct the radar record for topographic variations. Results of a walk-away test for velocity and dielectric permittivity were used to transform the radar time section to a depth section.

Ground-penetrating radar appears to work best in homogeneous, damp to wet, clay-free sediment (Olhoeft, 1990b). Several types of signal loss and attenuation can occur with radar waves. The most important losses for sand-sized sediment appear to be related to the presence of saline pore waters, and surface-reactive clay such as smectite in the interstices of the sand and in layers or drapes. Wave-

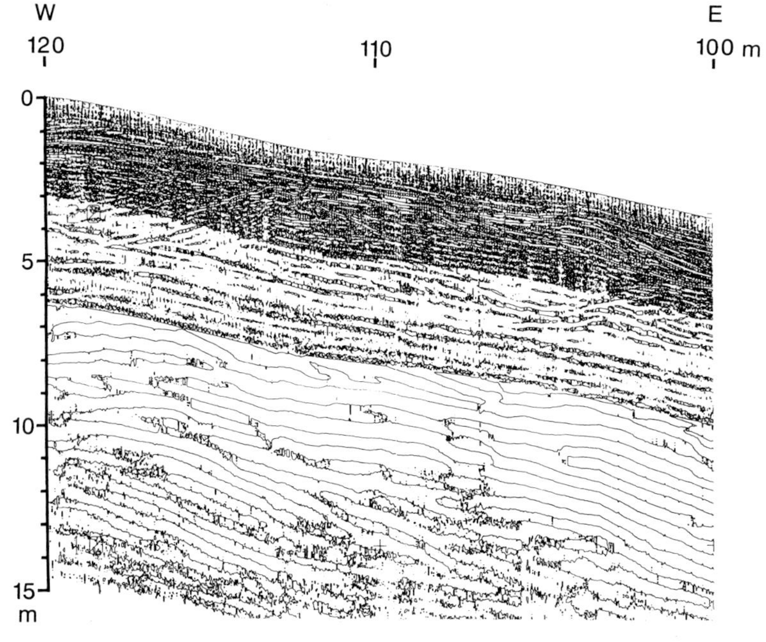

Fig. 1. Portion of the corrected radar depth section from 120 m to 100 m along the east–west transect (see Fig. 5), illustrating the radar record prior to the placement of interpretive lines. The banding (boundaries at 3 and 6.5 m) is due to the stacking of radar antenna records of differing frequencies and penetration efficacies. Note that reflectors clearly define sediment packages in the upper 5 m compared to the lower portion of the record; resolution falls rapidly at depths greater than about 5 m. Vertical and horizontal scales are in metres with no vertical exaggeration.

spreading losses can also be important if the sediment package in the sand body is predominantly on the same scale as the wavelength of the emitted radar energy.

RADAR SURVEY OF A LARGE DUNE

Description of study area

A large, complex aeolian dune situated along the eastern margin of Great Sand Dunes National Monument, Colorado, was chosen for the radar study (Fig. 2). Great Sand Dunes formed in a re-entrant along the margin of the Sangre de Cristo Mountains, along the eastern edge of the San Luis Valley (Wiegand, 1977). Sand-moving winds are predominantly from the southwest throughout most of the year, but strong easterly storm winds and minor winds from the north also result in sand transport. The upper parts of the large dunes are nearly reversed from the normal direction of migration during the strong east winds. The result of these multidirectional winds has been the accumulation of more than 250 m of aeolian dune sand along the eastern margin of the San Luis basin.

A comparison of sand-carrying wind directions with the orientation of the larger dunes (Fig. 2) suggests that the larger dunes are somewhat oblique to the dominant southwest winds, and nearly transverse, but in a reversed sense, to the strong, but short-lived east winds. The strongest control on maintenance of the large dune alignment may be a function of the east winds, but the major slip faces throughout the year are controlled by the southwest winds. The obliquity of the southwest winds to the large dunes may be partially responsible for the formation and migration of smaller dunes on the windward and lee slopes of the larger dunes. The smaller dunes are more mobile than the larger dunes.

The dunes along the eastern margin of the dunefield exhibit complex morphologies. The large dunes have slip faces whose orientation is partially maintained by the prevailing southwest wind. The upper parts of these slip faces are periodically blown back by strong east winds. In winter, strong winds associated with fronts commonly scour down to the level of damp sand (Fryberger, 1990a), in some areas removing tens of centimetres of sand. These winds serve to smooth or round off the upper parts of the dunes (Fig. 3). Many of the larger dunes have smaller dunes (<2 m high) superimposed on the windward and lee slopes. Strong winter winds commonly scour these small dunes down to low-relief features, forming low-angle erosional bounding surfaces. Strong east winds commonly form 'vortex scours' (Fryberger *et al.*, 1984) on the upper parts of the dunes during times of wind reversal. Vortex scours are wide, trough-shaped erosional depressions (outlined by wide, trough-shaped bounding surfaces) that can fill in during sand movement by the prevailing winds.

The general shape and location of these large dunes at Great Sand Dunes has remained relatively stable over a period of several years (Wiegand, 1977; Andrews, 1981), but the morphology and location of small-scale superimposed dunes change considerably from season to season.

Description of radar transect

The radar transect run during this study paralleled the predominant storm wind direction, and essentially followed an east–west orientation. The transect, approximately 300 m long, began in the bed of Medano Creek, proceeded west across flat alluvium, climbed the lee slope of the dune complex, and continued across the dune for the remainder of the transect. Stations were flagged every 10 m for location marks on the record, and were flagged as stations for the topographic profiling. Each antenna was pulled slowly along the transect (Fig. 4), and the data were recorded on tape for later processing. The portion of the 300 m transect illustrated in Fig. 5 (facing p. 64) of this study is the western 180 m that contained the aeolian dune sediments.

Results of the radar transect

The stacked radar cross-section illustrates the general attitude of reflectors to a depth of approximately 15 m along the east–west transect (see Fig. 5). Resolution of subsurface structure is best with the higher-frequency wavelengths, and resolution decreases with depth. The reflectors are the result of recorded differences in dielectric properties of the sediment, including changes in moisture and density. Solid (interpretive) lines on the radar cross-section represent boundaries between sediment packages where a distinct change in the attitude of bedding was interpreted from the reflectors. These solid lines outline erosional bounding surfaces between aeolian sediment packages. The loss of resolution with depth allows fewer interpretations of bounding surfaces in the deeper (>5 m) part of the

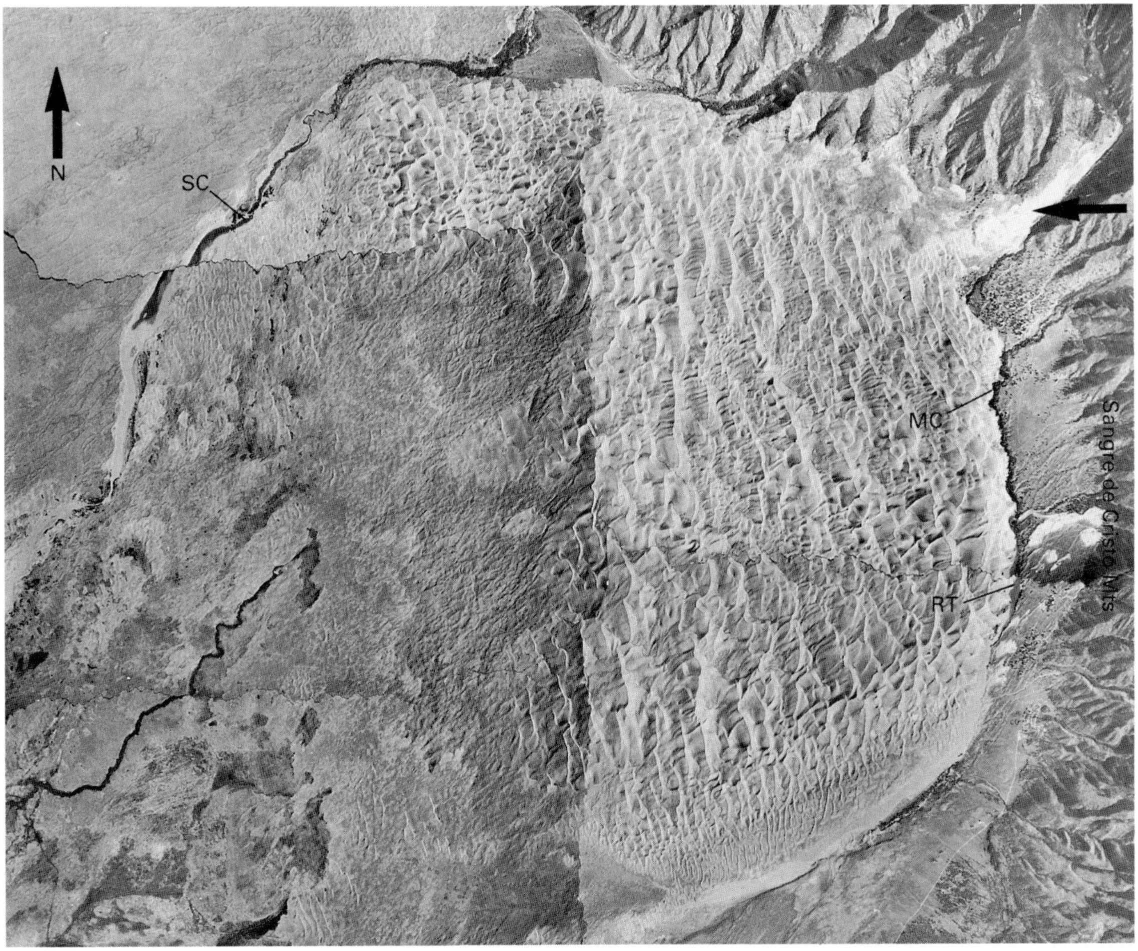

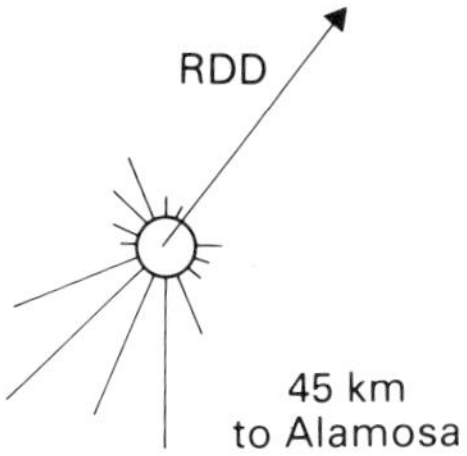

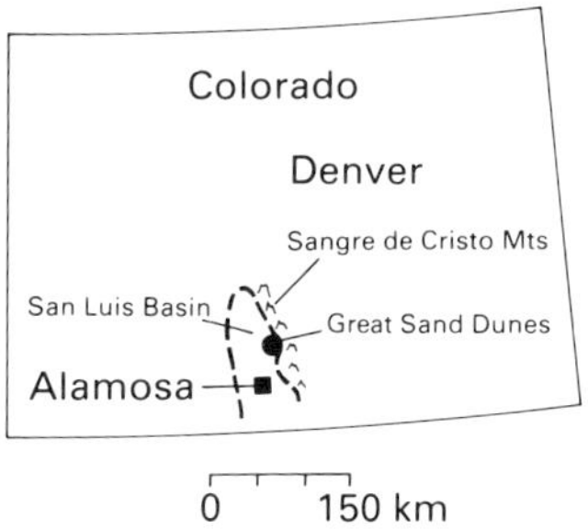

Fig. 2. Low-altitude aerial photomosaic of Great Sand Dunes National Monument, Colorado. The main dune mass is bordered on the east by Medano Creek (MC), and on the west by Sand Creek (SC). Strong east winds (black arrow) associated with storms cause significant sand transport and modification of the upper parts of the dunes and lee slopes. Predominant southwest winds are illustrated in a sand rose calculated from wind data collected near Alamosa, Colorado, 45 km southwest of Great Sand Dunes (from Fryberger *et al.*, 1979). The Alamosa sand rose does not record the capacity of the east winds to transport sand because of its distance from the dunes. The radar transect (RT) was situated on a large dune along the eastern margin of the dunefield. The white bar scale equals 1 km.

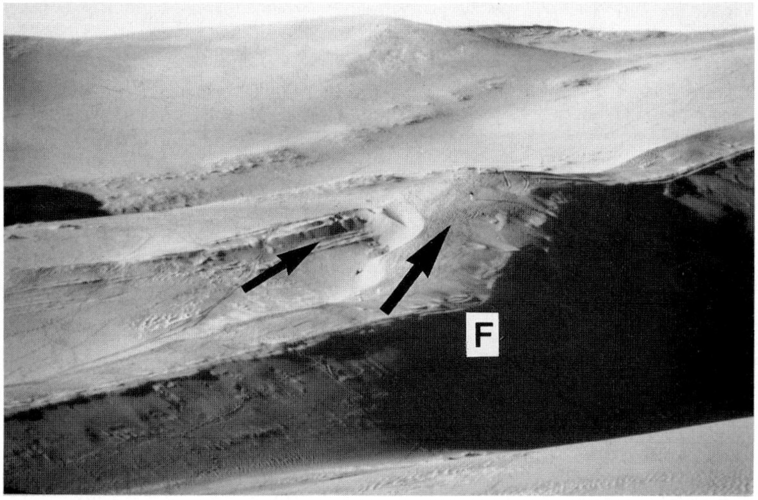

Fig. 3. Strong east winds are oblique to the predominant southwest winds, and modify the foreset slopes of the dunes. The upper part of the foreset (F) of this large dune has been blown back (large arrow) onto the windward slope of the dune by the east winds. These winds have also scoured the foreset slope down to damp sand, producing a subtle erosional bounding surface (reactivation surface) on the foreset. These bounding surfaces are common in the radar record (see Fig. 5, 70–20 m). Scouring has also exhumed damp sand strata on other parts of the dune (small arrow). Southwest winds will eventually remove this sand and deposit it on the foreset. The foreset on the bottom left-hand side of the figure is about 16 m high.

record. The dashed (interpretive) lines on the cross-section indicate reflectors that may represent either a selected set of cross-strata or the general attitude of a sediment package. Greater resolution in the upper 5 m of the cross-section meant that more bounding surfaces were interpreted in that portion.

Analysis of bounding surfaces and attitudes of reflectors in the cross-section indicates that this large dune, which is probably typical of dunes along the eastern margin of the dunefield, had a complex history of deposition and erosion. From 176 m to 150 m along the transect (Fig. 5), the reflectors define a large dune with foresets up to 23 m in length. The set of cross-strata ranges from 5 to 8 m in thickness. The sweeping, tangential nature of the reflectors indicates that a large dune foreset and dune apron with aeolian ripple strata were preserved. This large set grades to the east at 150 m into a zone of complex, irregular reflectors that extend to approximately 115 m. The loss of resolution with the 300 MHz antenna makes it difficult to interpret structures or bounding surfaces beyond the presence of irregular reflectors. These reflectors may represent sediment packages that were deposited by smaller dunes migrating with a north–south component. From 115 m to 20 m along the transect, the reflectors suggest that a large dune migrated to the east, possibly coeval with the foreset at 176–150 m. The reflectors outline sloping, possibly foreset, bedding that developed as the lee slope of the dune migrated to the east-northeast.

From 50 m to 20 m the higher-resolution antennas recorded reflectors that are interpreted as subtle erosional bounding surfaces formed on the lee slope of the dune. This portion of the transect is characterized by numerous bounding surfaces that partially bevel pre-existing foreset cross-strata. Such bounding surfaces may be common in the section from 115 m to 50 m, but the lack of resolution with the lower-frequency antenna (300 MHz) precludes their interpretation.

From 160 m to 80 m, the upper 3–5 m of the cross-section is characterized by numerous reflectors interpreted as trough-shaped bounding surfaces. The troughs are generally wide and shallow, ranging from 5 to 22 m in width, averaging 10 m, and ranging from 0.5 to 3 m in thickness, averaging 1 m.

From 40 m to 0 m on the transect, a series of strong reflectors that begin at the surface of Medano Creek gently dips beneath the dune sediments. This set of reflectors is interpreted as the expression of the local water table (outlined by the heavy black line). The radar transect was run in December; the creek was frozen, as was the upper 25 cm of the sand.

Evolution of dune structures

The western end of the transect begins with large dune foresets more than 20 m long migrating to the east-northeast, probably under the influence of the southwest winds. This dune foreset was probably active at the same time as the large foreset at 50–20 m. The zone of irregular reflectors between

Fig. 4. Radar operation. (A) View of the line of the radar transect from the bed of Medano Creek westward up the lee slope and across the large dune. The transect was measured by tape, and each 10 m segment was flagged as a station during each radar run (arrow). The transect was 300 m long, 180 m of which was across the large dune. The cable shown leads from the instrumentation to the antenna which was dragged across the sand. The person in the background gives the scale. (B) Typical radar run using the 900 MHz antenna (in small box on sand surface). The antenna is dragged at a constant speed across the sand. The handle contains an electronic switch that allows the user to place a timing mark on the radar record at each 10 m station. The person handling the cable enables the radar operator to pull the antenna at a constant speed without cable drag. The view is to the north. (C) The surface of the large dune in the study area is covered with scour features of low relief (a few tens of centimetres of relief) that were formerly active superimposed dunes. The dark areas are exposed damp sand; the remainder of the surface is covered with a thin (<2 cm) veneer of dry sand. The scour of the superimposed dunes produced a low-angle, subtle bounding surface. (D) The lee slope of the large dune in the eastern part of the study area was characterized by low-relief, scoured, superimposed dunes. At the time of the transect the dunes were scoured down to damp sand, producing a low-angle, subtle bounding surface on the lee slope (see Figs 3 and 5).

150 m and 115 m probably represents a common geomorphic situation along the eastern margin of Great Sand Dunes, in which smaller superimposed dunes migrate at some oblique angle to the migration direction of the larger dunes (Fig. 4D). For some reason the large dune at 176–150 m ceased to migrate, and probably devolved to the east into a low-angle slope covered by smaller superimposed dunes, the surface being affected by blowouts formed during wind reversals. Thus, the western dune gradually merged with the eastern dune, producing a larger complex dune that was partially covered with smaller dunes.

The multiple, subtle bounding surfaces on the leading edge of the dune are probably due to scouring by winds during reversals, and are of the type commonly called 'reactivation surfaces'. Reversing winds are typically related to storms, and strong winds associated with the storms scour down to the level of damp sand, in some cases several tens of centimetres deep. These scours are recorded by the bounding surfaces. During wind reversal, winds commonly travel at some oblique angle to the lee slope of the dune, scouring and depositing the sand on the windward slope of the dune (Fig. 3).

The zone of troughs between 160 m and 80 m may

have been formed by any of several mechanisms. Trough-shaped scour surfaces can form as pits in front of advancing bedforms, possibly by the migration of smaller dunes. Scours can also form in areas like Great Sand Dunes where wind reversals are common. The reversing winds impact the foresets on the dunes at an oblique angle; the wind commonly scours pits (vortex scours) into the upper parts of the dune foreset and the windward slope as strong winds reverse the normal direction of sand transport. The deep trough structure between 125 m and 110 m exhibits cross-strata facing away from the direction of the normal winds, probably filled with sand carried by reversing winds from the east. Other troughs in this zone may have had a similar origin. In some areas of Great Sand Dunes, winds scour broad troughs that are unrelated to either scours in front of dunes or vortex scours. This could be the origin for some of the troughs. Wide, trough-like structures have been described from other aeolian sands (McKee, 1966, 1979; Bigarella, 1979; Rubin & Hunter, 1983; Rubin, 1987).

Quantification of length-scales from the radar cross-section

Length-scales of sediment packages in two dimensions were measured from the radar cross-section. The troughs in the zone between 160 m and 80 m range from 5 to 22 m in width, averaging 10 m. The thickness of the troughs ranges from 0.5 to 3 m, and averages 1.1 m. The direction from which the troughs filled is variable. Some troughs appear to be filled with sand carried by east winds, some by southwest winds, and some by winds with a north–south component.

The subtle bounding surfaces at the leading edge of the dune bracket sets that range in length from 15 to 28 m, averaging 20 m, and range in thickness from 0.5 to 2 m, averaging 1 m. This scale of set geometry may be typical of a large part of the dune sand body from 90 m to 0 m along the cross-section. The one large set at the western end of the transect is 23 m long, and ranges from 5 to 8 m thick.

Other miscellaneous sets in the cross-section with reliable bounding surfaces range from 6 to 12 m in length, averaging 8.5 m, and range in thickness from 0.75 to 1.5 m, averaging 1 m.

The data on set length and thickness demonstrate the variability in set dimensions within one large dune at Great Sand Dunes. The effect of the reversing east wind is important in producing conditions of erosion that lead to the formation of the trough-shaped sets and to the numerous reactivation surfaces along the leading edge of the dune. These erosional bounding surfaces are preserved, and result in the sand body being comprised of numerous sets with differing geometries. This large dune at Great Sand Dunes is likely to be partially preserved, as it is being covered by the next large, climbing dune.

DISCUSSION

The radar-derived cross-section illustrates the relationship between the complexity of dune internal structure and the complexity of sand-transporting winds. Winds from several directions can scour as well as deposit sand during different times of the year, creating bounding surfaces and deposits that would not be predicted to occur from the external morphology of the dune alone. Dunes with simple external morphologies can exhibit complex internal structures. McKee (1966) demonstrated the presence of numerous bounding surfaces in externally simple dunes at White Sands, New Mexico. The wind rose from White Sands indicates that sand-transporting winds are multidirectional (Fryberger, 1979), suggesting a relationship between wind complexity and complexity of dune structure. Dune type may be a determinant of the degree of complexity of internal structure, as barchanoid dunes at White Sands had more complex internal structures than the transverse dune. Aeolian dune type is strongly controlled by local wind patterns (Fryberger, 1979; Lancaster *et al.*, 1987).

Coastal dunes of Oregon are constructed by sand transported from several directions in a seasonally controlled pattern (Hunter *et al.*, 1983). Numerous bounding surfaces and cosets of strata are produced in the dunes by the directional variations in sand-transporting winds, resulting in complex internal structures (Fryberger, 1990b). These examples indicate that a fruitful avenue of research would be to examine in more detail the relationship between wind complexity and complexity of internal structure in modern dunes. Only this way will progress be made in applying knowledge of modern dune structures to understanding complex structures in aeolian sandstones. The careful use and interpretation of radar data may comprise a valuable tool to investigate this relationship.

Aeolian sandstones can exhibit complex internal architecture. Rubin (1987) demonstrated with numerous examples from aeolian sandstones the complexities of internal structure produced by migrating aeolian bedforms. He interprets palaeodune morphology and general palaeowind regime from a careful analysis of bounding surfaces and aeolian coset architecture. Radar-derived architectural data from modern aeolian dunes formed in known wind regimes may help refine input data for complex bedform migration models such as those he has created.

The complexity of dune internal structure is also important in understanding and predicting types of fluid-flow heterogeneity in aeolian hydrocarbon reservoirs. Aeolian sandstones produce oil and gas in many areas, and the reservoirs are in part heterogeneous because of internal architecture (Weber, 1987; Lindquist, 1988; Luthi & Banavar, 1988). Data such as characteristic cross-strata set dimensions are needed by engineers as they attempt to increase fluid recoveries from aeolian reservoirs.

SUMMARY AND CONCLUSIONS

The ground-penetrating radar recorded reflectors that define bounding surfaces and internal sediment packages in the upper 5 m of a complex aeolian dune along the margin of Great Sand Dunes. The radar resolution at depths greater than 5 m was generally not enough to resolve bounding surfaces of any variety with certainty. The radar demonstrated that dunes which are seemingly simple from the viewpoint of exterior morphology can have complex internal structures. Work is currently in progress on determining the three-dimensional length-scales of internal sedimentary units in these dunes.

The radar worked well in damp to dry aeolian sand at Great Sand Dunes. The radar apparently does not work well in sediment with a high percentage of surface-reactive clays such as smectite, or where pore waters are saline. These conditions may limit its usefulness in delineating sedimentary structures in other sedimentary environments.

ACKNOWLEDGEMENTS

We thank William Wellman, Superintendent of Great Sand Dunes National Monument, Colorado, for his unfailing support of our radar work, and for his support of science in general at Great Sand Dunes for the past several years. William Havilland and Fred Bunch of the National Park Service were invaluable to the study, as was Edgar Ethington of the USGS.

REFERENCES

Andrews, S. (1981) Sedimentology of Great Sand Dunes, Colorado. In: *Recent and Ancient Non-marine Depositional Environments: Models for Exploration* (Ed. Ethridge, F.G. & Flores, R.M.) pp. 279–291. Spec. Publ. Soc. Econ. Paleontol. Mineral., Tulsa, 31.

Bigarella, J.J. (1979) Structural features at Lagoa dune field, Brazil. In: *A Study of Global Sand Seas* (Ed. McKee, E.D.) pp. 114–134. US Geol. Surv. Prof. Pap. 1052.

Brookfield, M.E. (1977) The origin of bounding surfaces in ancient aeolian sandstones. *Sedimentology* **24**, 303–332.

Daniels, J.J., Vendl, M. & Stepp, T. (1988) Ground penetrating radar investigations of near-surface contaminant spills. *Soc. Explor. Geophys., 58th Annual Meeting, Denver, Abstract No. 3.5*, 343–344.

Duke, S.K. (1990) *Calibration of ground-penetrating radar and calculation of attenuation and dielectric permittivity versus depth*. Unpublished Master's thesis, Colorado School of Mines, Golden.

Fryberger, S.G. (1979) Dune form and wind regime. In: *A Study of Global Sand Seas* (Ed. McKee, E.D.) pp. 137–169. US Geol. Surv. Prof. Pap. 1052.

Fryberger, S.G. (1990a) Great Sand Dunes depositional system — an overview. In: *Modern and Ancient Eolian Deposits: Petroleum Exploration and Production: Rocky Mount* (Fryberger, S.G., Krystinik, L.F. & Schenk, C.J.) pp. 1-1–1-9. Rocky Mtn Sec. Soc. Econ. Paleontol. Mineral., Denver.

Fryberger, S.G. (1990b) Coastal deposits of Oregon, USA, Guerrero Negro, Mexico, and Jafurah sand sea, Saudi Arabia. In: *Modern and Ancient Eolian Deposits: Petroleum Exploration and Production: Rocky Mount* (Fryberger, S.G., Krystinik, L.F. & Schenk, C.J.) pp. 11-1–11-15. Rocky Mtn Sec. Soc. Econ. Paleontol. Mineral., Denver.

Fryberger, S.G. (1991) Personal communication.

Fryberger, S.G., Ahlbrandt, T.S. & Andrews, S. (1979) Origin, sedimentary features, and significance of low-angle eolian 'sand sheet' deposits, Great Sand Dunes National Monument and vicinity, Colorado. *J. Sed. Petrol.* **49**, 733–746.

Fryberger, S.G., Al-Sari, A.M., Clisham, T.J., Rizvi, S.A.R. & Al-Hinai, K.G. (1984) Wind sedimentation in the Jafurah sand sea, Saudi Arabia. *Sedimentology* **31**, 413–431.

Hunter, R.E., Richmond, B.M. & Alpha, T.R. (1983) Storm-controlled oblique dunes of the Oregon coast. *Bull. Geol. Soc. Am.* **94**, 1450–1465.

Imai, T., Sakayama, T. & Kanemori, T. (1987) Use of ground-probing radar and resistivity surveys for archaeological investigations. *Geophysics* **52**, 137–150.

Kocurek, G. (1988) First-order and super bounding surfaces in eolian sequences — bounding surfaces revisited. *Sed. Geol.* **56**, 193–206.

Kocurek, G. (1991) Interpretation of ancient eolian sand dunes. *Ann. Rev. Earth Planet. Sci.* **19**, 43–75.

Lancaster, N., Greeley, R. & Christensen, P.R. (1987) Dunes of the Gran Desierto sand-sea, Sonora, Mexico. *Earth Surf. Proc. Landf.* **12**, 277–288.

Lindquist, S.J. (1988) Practical characterization of eolian reservoirs for development: Nugget Sandstone, Utah–Wyoming thrust belt. *Sed. Geol.* **56**, 315–339.

Luthi, S.M. & Banavar, J.R. (1988) Application of borehole images to three-dimensional modeling of eolian sandstone reservoirs, Permian Rotliegende, North Sea. *Bull. Am. Assoc. Petrol. Geol.* **72**, 1074–1089.

McKee, E.D. (1966) Structures of dunes at White Sands National Monument, New Mexico (and a comparison with structures of dunes from other selected areas). *Sedimentology* **7**, 1–69.

McKee, E.D. (1979) Sedimentary structures in dunes. In: *A Study of Global Sand Seas* (Ed. McKee, E.D.) pp. 83–113. US Geol. Surv. Prof. Pap. 1052.

Moorman, B.J., Smith, D.D., Judge, A.S. & Jol, H.M. (1990) Assessing the ability of ground penetrating radar to delineate different fluvial systems [abstract]. In: *Third International Conference on Ground Penetrating Radar; Abstracts of the Technical Meeting; Lakewood, Colorado* (Ed. Lucius, J.E., Olhoeft, G.R. & Duke, S.K.) p. 44. US Geol. Surv. Open-file Rep. 90-414.

Olhoeft, G.R. (1990a) Ground penetrating radar data processing and modeling [abstract]. In: *Third International Conference on Ground Penetrating Radar; Abstracts of the Technical Meeting; Lakewood, Colorado* (Ed. Lucius, J.E., Olhoeft, G.R. & Duke, S.K.) p. 49. US Geol. Surv. Open-file Rep. 90-414.

Olhoeft, G.R. (1990b) Personal communication.

Rubin, D.M. (1987) Cross-bedding, bedforms, and paleocurrents. *Soc. Econ. Paleontol. Mineral., Concepts in Sedimentology and Paleontology 1*. SEPM, Tulsa, 187 pp.

Rubin, D.M. & Hunter, R.E. (1983) Reconstructing bedform assemblages from compound cross-bedding. In: *Eolian Sediments and Processes* (Ed. Brookfield, M.E. & Ahlbrandt, T.S.) pp. 407–427. Developments in Sedimentology 38. Elsevier, Amsterdam.

Weber, K.J. (1986) How heterogeneity affects oil recovery. In: *Reservoir Characterization* (Ed. Lake, L.W. & Carroll, H.B.) pp. 487–544. Academic Press, New York.

Weber, K.J. (1987) Computation of initial well productivities in aeolian sandstone on the basis of a geological model, Leman Gas Field, UK. In: *Reservoir Sedimentology* (Ed. Tillman, R.W. & Weber, K.J.) pp. 333–354. Spec. Publ. Soc. Econ. Paleontol. Mineral. 40. SEPM, Tulsa.

Wiegand, J.P. (1977) *Dune morphology and sedimentology at Great Sand Dunes National Monument, Colorado*. Unpublished Master's thesis, Colorado State University, Ft Collins.

Spec. Publs Int. Ass. Sediment. (1993) **16**, 71–83

Origins and sedimentary features of supersurfaces in the northwestern Gran Desierto Sand Sea

N. LANCASTER

Quaternary Sciences Center, Desert Research Institute, University of Nevada System, PO Box 60220, Reno, NV 89506, USA

ABSTRACT

Studies of sand sheets and dunes in the northwestern margins of the Gran Desierto Sand Sea provide criteria for the recognition of regional-scale or super bounding surfaces in aeolian sequences. Genetic units in the sand sheets and dunes are separated by supersurfaces. The sand sheets were formed in conditions of limited sand supply, sediment bypassing and a sparse vegetation cover. Termination of sand supply to the area from the Colorado River valley resulted in deflation of sand sheet deposits until a protective lag had formed on supersurface A. Renewed sand sheet and dune accumulation was ended by climatic changes resulting in an increase in the vegetation cover, stabilization of the surface and pedogenic alteration of underlying deposits to form supersurface B. In this area, it appears that supersurfaces resulting from termination of sand supply are characterized by deflation and development of surface lags, whereas those resulting from regional climatic changes are marked by pedogenic alteration.

INTRODUCTION

The processes by which large aeolian sand bodies (sand seas, ergs) accumulate are imperfectly known. Recently, a number of workers (e.g. Kocurek, 1988; Havholm *et al.*, this volume) have recognized regional-scale bounding surfaces (supersurfaces) in some ancient aeolian sandstones and shown that they separate genetically distinct sand bodies. The surfaces represent hiatuses of erosion or non-accumulation in aeolian sequences which resulted from changes in sea level, climate, regional tectonics or sand sea migration. In modern sand seas, there is an increasing body of evidence that suggests that they too have accumulated episodically (e.g. Talbot, 1985; Kocurek *et al.*, 1991; Lancaster, 1990), with Quaternary climatic and eustatic changes providing a major control on aeolian sedimentation patterns.

Recognition and tracing of super bounding surfaces make up a potentially important tool for regional stratigraphic correlation of aeolian sandstones, as well as for the reconstruction of the depositional history of ancient and modern sand seas. In order to be able to reconstruct depositional histories effectively, it is important that supersurfaces formed by different sets of processes can be recognized at the outcrop scale. Studies of supersurfaces formed in modern sand seas can help to provide criteria to identify the processes that created ancient supersurfaces. This paper documents (i) the existence of supersurfaces in the northwestern part of the Gran Desierto Sand Sea in northern Mexico, and (ii) the sedimentary features associated with these surfaces. Possible mechanisms for the formation of the surfaces are then discussed.

THE GRAN DESIERTO SAND SEA

The Gran Desierto Sand Sea (Fig. 1) covers an area of 5700 km^2 on the northeastern shores of the Gulf of California, east of the Colorado River delta. There are two main areas of dunes in the sand sea (Lancaster *et al.*, 1987): an eastern group of compound and complex crescentic dunes 10–80 m high, and a western area of chains and clusters of star dunes 80–150 m high that are surrounded by reversing and crescentic dunes 10–20 m high

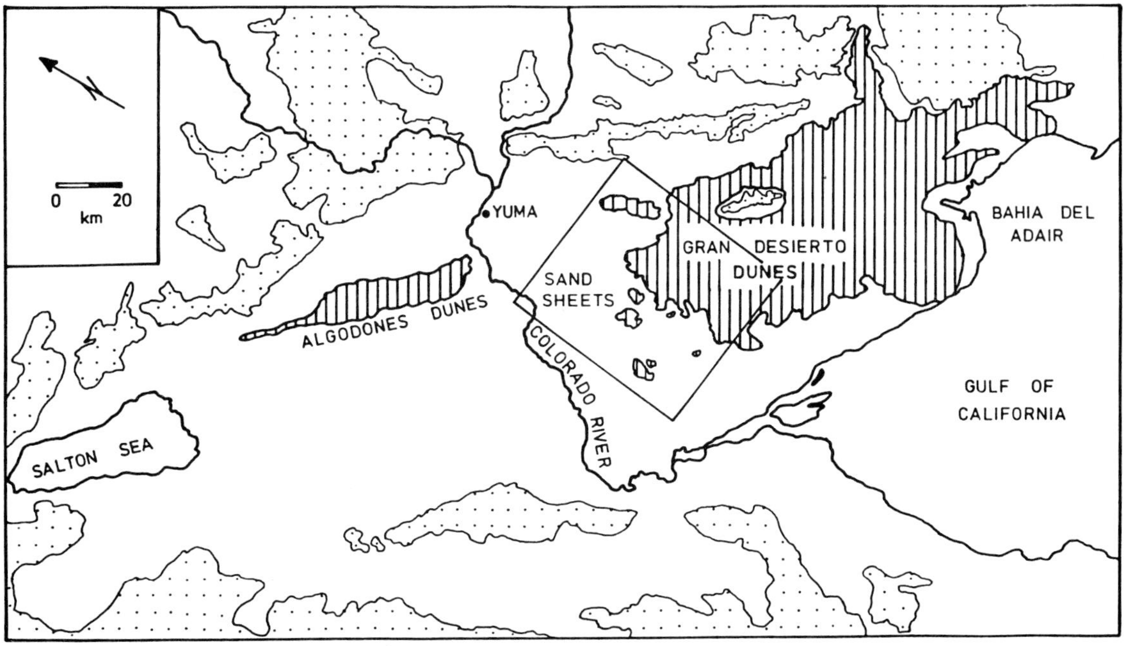

Fig. 1. The Gran Desierto Sand Sea: location map. Box indicates area covered by Landsat image in Fig. 2.

(Fig. 2). In addition, there are areas of stabilized crescentic dunes 5–10 m high on the margins of the star dune area and in some inter-dune areas between the chains of star dunes. Sand sheets and streaks and small areas of active crescentic dunes occur in the west and northwest parts of the sand sea. The western star dunes and the eastern crescentic dunes are separated by an undulating vegetated sand sheet

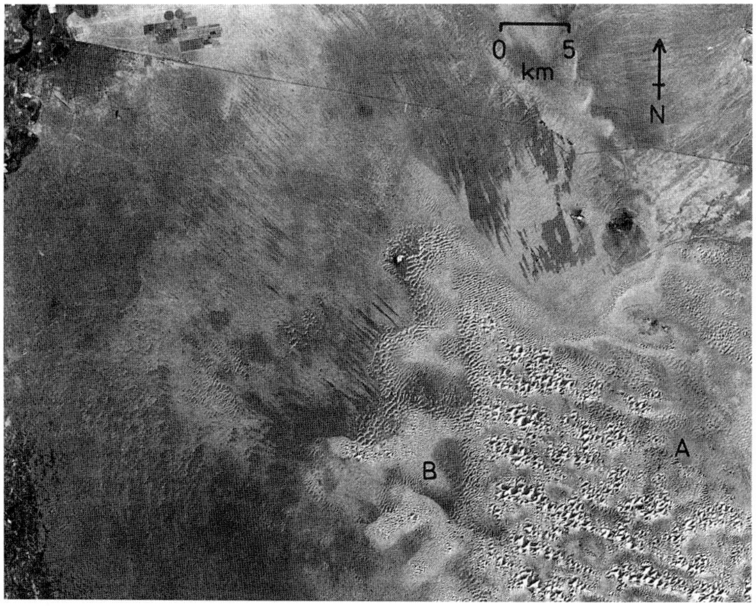

Fig. 2. Landsat image of northwestern part of Gran Desierto Sand Sea. Note areas of star dune chains to southeast, stabilized crescentic dunes (A), crescentic and reversing dunes on margin of sand sea (B), and sand sheets and streaks west of these dunes.

and small areas of active crescentic dunes.

The Gran Desierto is an excellent area in which to study the effects of climatic, tectonic and eustatic changes on sand supply and sand sea development. The area is situated on the margins of a rapidly subsiding basin, the Salton Trough. The delta region of the Colorado River has been deformed extensively over the past 5.5m yrs in association with the opening of the Gulf of California (Winkler & Kidwell, 1986), resulting in a westward migration of the Colorado River during the Pliocene and Pleistocene and uplift of the Mesa Arenosa along the western margin of the sand sea (Colletta & Ortlieb, 1984). Evidence from lake levels and pack rat middens (Van Devender *et al.*, 1987; Van Devender, 1990) indicates that the region that includes the Gran Desierto Sand Sea has experienced a number of periods of increased rainfall during the late Pleistocene and Holocene, although the area of the sand sea has probably always been arid during this time (Cole, 1986). Eustatic changes in sea level in the region have had a major influence on the pattern of deltaic sedimentation (van Andel, 1964). As the Colorado delta is a major source of sediment for the sand sea, these events probably had a significant influence on aeolian sedimentation patterns.

The Gran Desierto Sand Sea has accumulated episodically since the mid-Pleistocene (Blount & Lancaster, 1990). Evidence from dune morphology and the texture and mineralogy of aeolian sediments indicates that there are three major sand populations, representing three pulses of sediment input. The oldest sands were derived from the ancestral Colorado River floodplain and delta. Later inputs of sand came from the modern Colorado River floodplain and coastal areas south of the sand sea in the region of the Bahia del Adair. Relatively brief periods of aeolian deposition have been separated by long intervals of geomorphic stability during which existing sand populations were reworked, soils were formed and deflation of existing dune areas and sand sheets occurred in many parts of the sand sea. The extensive sand sheets of the Gran Desierto that are juxtaposed with dunes of apparently different ages and source sediments provide an excellent opportunity to examine supersurfaces in a modern aeolian environment.

SAND SHEETS IN THE NORTHWESTERN GRAN DESIERTO SAND SEA

Topography and general description

Sand sheets cover an area of 1000–1500 km^2 in the northwestern parts of the Gran Desierto (Fig. 2). They form a flat to gently undulating surface with a maximum local relief of 1–5 m (Fig. 3), which includes several small areas of very degraded, stabilized crescentic dunes 3–5 m high (A on Fig. 2). Vegetation consists mostly of sparse grass clumps and creosote bushes. To the north and west, the sand sheets thin out on the highest alluvial terraces of the Colorado River and the distal areas of alluvial fans that head in the Tinajas Altas Mountains. To

Fig. 3. Ground view of sand sheets. View northeast towards Tinajas Altas Mountains (on horizon). Vegetation is mostly dead white bursage (*Ambrosia dumosa*).

the southeast, the sand sheets are intercalated with crescentic and reversing dunes that form the western margin of the main dune area. In many respects, the area appears to be a typical 'back erg' (Porter, 1986) or 'trailing erg margin' (Sweet *et al.*, 1988).

Landsat images of the region (Fig. 2), together with field investigations, show that the sand sheets consist of a series of coalescing streaks or stringers aligned at 150° in central and western parts of the area, and 165° in the east (Fig. 2). The streaks are imaged as areas of higher albedo that contrast with a darker underlying unit, the surface of which is composed of a coarse sand and granule lag.

Trenching and sampling of the sand sheets show that they are composite features consisting of successive generations of aeolian accumulations. Three main genetic units can be recognized, termed Units I–III in order of increasing age. They are separated by two prominent and laterally extensive surfaces (Surfaces A and B). Figure 4 shows the topographic and stratigraphic relationships between the units and surfaces in representative areas of the sand sheets.

Description of sand sheet units

Unit I

Unit I forms the surface sands over large areas of the sand sheets. It ranges in thickness from as little as 5 cm in hollows to 1.5 m on the crests of the higher undulations, with an average thickness of 20–40 cm. Below the active wind ripples of the surface, Unit I consists of well developed horizontal to gently dipping (2–3° apparent dip) planar sets of wind-ripple laminae (Fig. 5). Very similar sedimentary structures have been described from sparsely vegetated sand sheets in Saudi Arabia and at Great Sand Dunes (Fryberger *et al.*, 1979, 1984; Kocurek & Nielson, 1986). In the eastern parts of the sand sheet, distinct sets of very fine (3.5–4.0 ϕ) and coarse (2 ϕ) laminae 1–5 cm thick equivalent to the 'Type a' of Fryberger *et al.* (1979) are common. There is very little evidence of bioturbation or reactivation surfaces within this unit.

Unit I is composed of light brown to reddish yellow (Munsell 7.5 YR 6/4–6/6) sand. The sand is

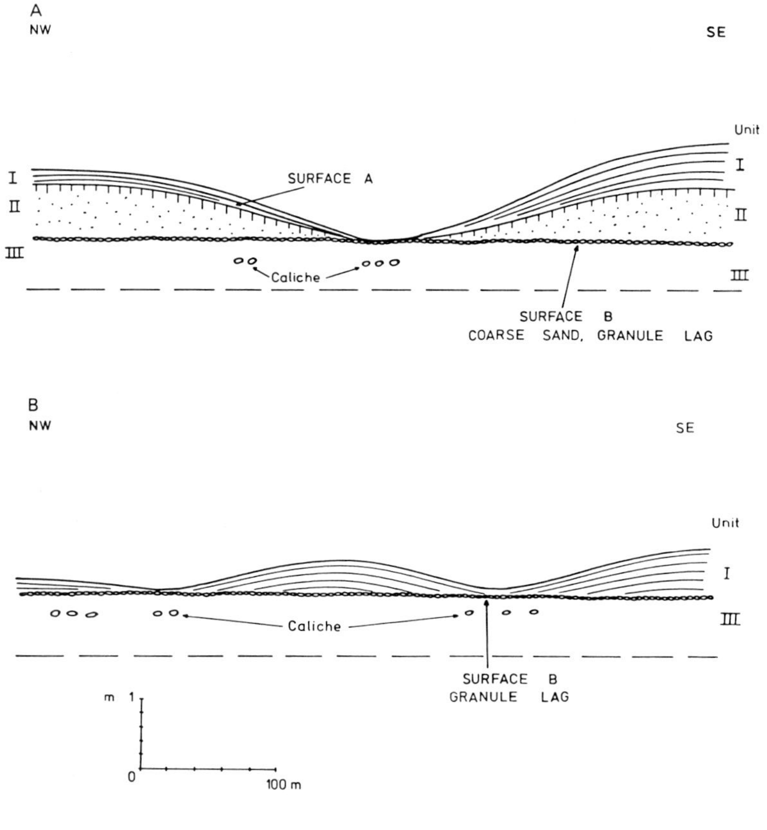

Fig. 4. Topographic and stratigraphic relationships between sedimentary units in the sand sheets. (A) In central areas of the sand sheets. (B) In northern areas.

Fig. 5. Sedimentary structures in Units I and II. Note parallel wind-ripple laminae of Unit I and structureless sands of Unit II.

predominantly (>90%) subrounded to subangular quartz. The fine and very fine sand fractions are clear or frosted. Medium and coarse sands commonly have a red brown (2.5 YR 5/6) patina. The coarse and very coarse sand fractions are dominated by dark lithic fragments and feldspars. The texture of Unit I is variable, but commonly consists of 40–50% medium sand and 23–45% fine sand (Fig. 6). Most deposits of Unit I are unimodal, with a modal size of 2.0 ϕ (0.25 mm). Mean grain size varies between 1.95 and 2.25 ϕ. Bulk samples exhibit moderate to poor sorting ($\sigma_I = 0.70-1.10$), although individual laminae appear well sorted. Samples from the eastern part of the sand sheet area are somewhat finer, with a mean grain size of 2.32 ϕ and moderate sorting ($\sigma_I = 0.81$). In this area, which is located close to areas of active crescentic dunes, Unit I sand is bimodal, with a coarse mode at 2.0 ϕ and a fine mode at 3.5–4.0 ϕ.

Unit II

Unit II underlies Unit I in the high albedo streaks, but not in the darker areas or in the northeast part of the area (Fig. 4A). In the coalescing streaks, Unit I is separated from Unit II by a horizon 5–10 cm thick in which the wind-ripple laminae become indistinct (Fig. 7). This zone forms a palaeosol that is capped by a weakly developed surface crust (vesicular A horizon). The palaeosol follows the local topography in a subdued manner, and can be traced throughout the higher areas of the sand sheets and into adjacent dunes. The laterally extensive nature of this feature suggests that it is a supersurface in the terminology of Talbot (1985) and Kocurek (1988) and is termed Surface A.

Unit II is composed of reddish yellow (7.5 YR 6/6), very weakly laminated to structureless, medium to fine quartz sand which is moderately to weakly indurated. The thickness of Unit II ranges from 20 cm on the flanks of the sand streaks to 1.5 m in the highest areas of the sand sheets. Unit II is texturally very similar to Unit I, and is composed of approximately equal proportions of medium and fine sand (Fig. 6B). The mode is at 2.0 ϕ. Mean grain size ranges from 2.10 to 2.50 ϕ and bulk samples are moderately to poorly sorted ($\sigma_I = 0.70-1.10$).

Unit III

Unit III directly underlies Unit II in most parts of the area. However, in the eastern area of the sand sheets, Unit III underlies Unit I (Fig. 4B). A sharp boundary with a well developed surface crust (vesicular A soil horizon) and a coarse sand to granule lag one or two grain diameters thick separates Unit III from overlying units (Fig. 8). This boundary (Surface B) appears to be planar in nature and can be traced to the dark streak areas where it is exposed at the surface (Fig. 9). Unit III is at least 1 m thick and overlies indurated silty sands and gravels of the ancestral Colorado River.

Unit III (Fig. 9) is composed of structureless, reddish yellow (7.5 YR 6/6), moderately consolidated

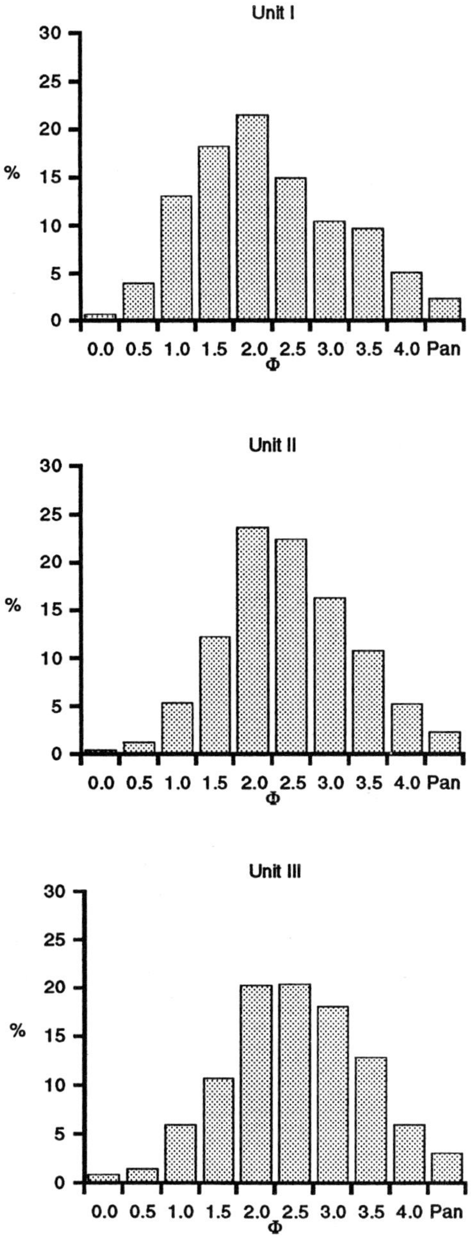

Fig. 6. Typical particle size distributions from sand sheet units.

fine to very fine sand, with rare gravel clasts. Compared to Units I and II, there is an increase in very fine sand to 15–20%, and a significant (3–5%) silt and clay content in this unit. The mode remains at 2.0 ϕ. Unit III is generally finer than other units, with a mean grain size of 2.10–2.45 ϕ, and is always moderately sorted ($\sigma_I = 0.75-0.90$). Calcium carbonate occurs in a diffuse form throughout the unit (total carbonate content = 2–5%), with small (1–3 cm) caliche nodules at a depth of 50–70 cm in some localities. In the classification of Gile *et al.* (1966), Unit III exhibits weak Stage II caliche development.

Dunes in the northwestern Gran Desierto

There are several sets of degraded and vegetated crescentic dunes with a height of 2–4 m and a spacing of 100–200 m within the area of the sand sheets (area A on Fig. 2). These dunes have stoss slope angles between 3 and 10°, and lee faces that slope at 10–20° to the southwest. In addition, vegetated and mostly stabilized crescentic dunes with a height of 5–10 m and a spacing of 300–500 m occur on the western margins of the main dune area of the sand sea (B on Fig. 2). In both examples, vegetation cover declines and the percentage of the surface covered by wind ripples increases up the stoss slopes, so that the broad flat crestal areas consist largely of unvegetated active sands. In all cases, these dunes overlie sand sheet Unit III and rest on Surface B (Fig. 10).

The strike of the crestlines of most of these dunes is 060–065°, indicating formation by winds from the northwest to north-northwest. This compares with a strike of 080–085° for currently active crescentic and reversing dunes on the western margins of the sand sea. The crestlines of these dunes are perpendicular to the gross bedform-normal vector (Rubin & Ikeda, 1990) of sand transport in the region, which is the resultant of winds from the north-northwest–north and south-southeast–south and suggest a shift in dune forming winds to the north in modern times.

Trenching of the degraded crescentic dunes shows that they consist of two main units (Fig. 10). Unit 1 is light yellowish brown to light brown (7.5 YR 6/4–10 YR 6/4) and consists of well developed wind-ripple laminae. It forms the active 'cap' of the dunes, and ranges in thickness from 5–10 cm on the lower stoss slopes of the dunes to as much as 2 m at the crest, declining again to 10–20 cm on the lee face. Sands of Unit 1 are typically coarse (mean grain size 1.25–2.00 ϕ), and moderately to poorly sorted ($\sigma_I = 0.70-1.10$). They are unlike the surface sands of the active dunes of the main sand sea to the southeast, which are commonly finer (mean grain size 2.05–2.45 ϕ), and well to very well sorted

Fig. 7. Surface A (arrowed) forming division between Units I and II.

Fig. 8. Surface B (arrowed) between Units I and III.

($\sigma_I = 0.33-0.57$), but are very similar to those of Unit I of the adjacent sand sheets.

Unit 2 of the dune sediments consists of structureless or very weakly laminated reddish brown (7.5 YR 6/6) medium to fine sands. There is extensive root mottling and bioturbation of this unit, especially on the lee face (Fig. 11). Unit 2 is much finer than Unit 1, with a mean grain size between 2.0 and 2.5 ϕ, and is better sorted ($\sigma_I = 0.60-0.80$). Texturally, Unit 2 is much closer to the crest sands of the active dunes of the sand sea. A distinct surface similar to Surface A in the sand sheets separates dune Unit 2 from Unit 1.

DISCUSSION

Origin of the sand sheets

Sand sheets are an important component of many sand seas and may form in conditions that include a high water table, periodic flooding, significant amounts of coarse sands, surface cementation and the presence of a vegetation cover (Kocurek & Nielson, 1986). Of these factors, coarse sand and/or vegetation seem to be the most common factors associated with sand sheets world-wide. Closely associated with areas of sand sheets in many sand

Fig. 9. Unit III. Note structureless character of unit, surface crust and lag of granules and very coarse sand.

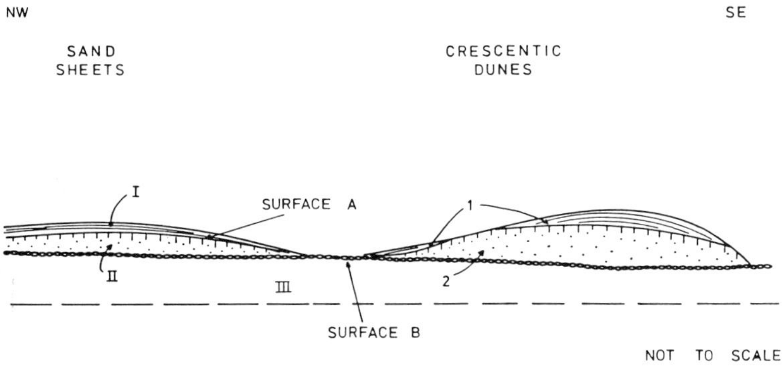

Fig. 10. Schematic view of dune stratigraphy and comparison with that of sand sheet areas.

seas are low-relief bedforms without slip faces called zibars (Nielson & Kocurek, 1986).

Although high water tables may have occurred locally in the northwestern Gran Desierto around Laguna Prieta, there is no evidence (e.g. in the form of secondary precipitation of evaporite minerals) to suggest that high water tables have been significant in the formation of the sand sheets in this area.

Comparisons of the composition of the sand sheets in the northwestern Gran Desierto with those in some other sand seas (Fig. 12) reveal that they lack the prominent coarse or very coarse sand fraction that is characteristic of many unvegetated or sparsely vegetated sand sheets. The modal size fraction in the Gran Desierto sand sheets is 2.0 or 2.5 ϕ (0.18–0.25 mm), compared to 0.5–1.5 ϕ (0.35–0.71 mm) in Namibian and Algodones sand sheets and zibars. With the exception of some Unit I sands, the Gran Desierto sand sheets are also unimodal with a prominent mode at 2ϕ, in contrast to many unvegetated sand sheets and zibars, which tend to be bimodal (Kocurek & Nielson, 1986). The active 'cap' of many crescentic dunes, as well as the active reversing and star dunes in the western part of the Gran Desierto Sand Sea, are composed of sand that has a mode of 2.0 or 2.5 ϕ, which is typical of many dune sands. The sand sheets in the northwestern parts of the Gran Desierto are therefore composed of sands that are clearly in the range commonly found in dunes, and their formation is not the result of a significant proportion of coarse or very coarse sand, which acts to suppress the formation of dunes with slip faces.

Today, the vegetation cover in the area is sparse, with a maximum percentage cover of 10–15%. This is insufficient to prevent sand transport taking place, but sufficient to cause divergence and convergence of airflow around individual plants in the manner

Fig. 11. Extensive bioturbation in Unit 2 of the dunes, overlain by weakly laminated Unit 1.

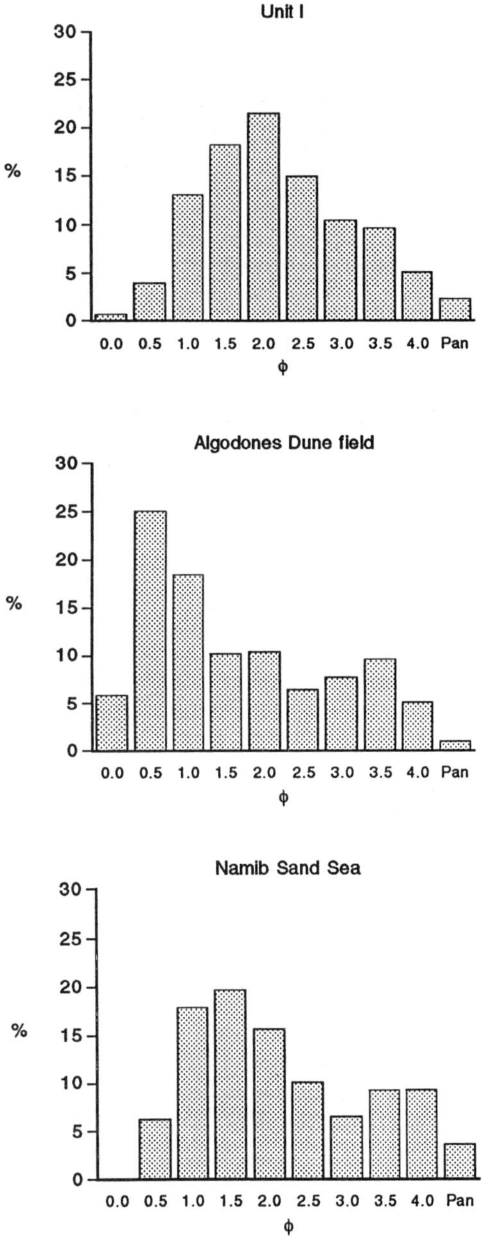

Fig. 12. Comparison between particle size distributions of Gran Desierto sand sheet Unit I and deposits of sand sheets in the Algodones dunefield and the Namib Sand Sea.

suggested by Ash & Wasson (1983) and Fryberger *et al.* (1979), and giving rise to localized deposition. Well sorted fine sand is deposited in 'sand tails' in the lee of grass clumps during periods of strong winds and covers pre-existing wind ripples (Fig. 13). Wind ripples can migrate into areas protected by vegetation, where local increases in aerodynamic roughness and decreases in near-surface wind speeds result in deposition (e.g. Hesp, 1983). If sand supply is very high then it is possible that these effects can lead to dune initiation (e.g. Kocurek *et al.*, 1992). However, dunes are not being initiated by these processes in this part of the Gran Desierto. It appears therefore that sand supply is a limiting factor, and the sparse vegetation cover leads to deposition of a poorly sorted mixture of fine and coarse sand. Sand sheet Unit I and dune Unit 1 were (and probably still are) being deposited under these conditions. In the terminology of Kocurek & Havholm (in press), the area today is probably a bypass supersurface, with a neutral sediment budget. Any accumulation such as Unit I therefore probably represents a temporary storage of medium and fine sand that is

Fig. 13. Deposition on sand sheets on the margins of the Gran Desierto star dune area. Note 'sand tails' or shadow dunes behind ephemeral herbs and isolated creosote bushes, and migrating wind ripples in intervening areas.

en route to its final depositional site in the dunes of the sand sea.

Units II and III in the sand sheets lack wind-ripple laminations, but have a texture that is similar to Unit I. They were probably deposited in similar conditions and originally had a similar set of sedimentary structures. These have been destroyed by post-depositional bioturbation and pedogenesis. The main effects of pedogenesis have been an increase in the percentage of very fine sand and silt and clay by infiltration of fines and possibly by weathering, but no Bt horizons or clay lamellae were noted. Some reddening has occurred, evidenced by an increase in Munsell chroma from /4 to /6. Unit III is weakly cemented by calcium carbonate. However, the degree of soil development in Unit II is limited compared to that described from dune areas in Australia (e.g. Churchward, 1963), the Sahel (Vökel, 1989), New Mexico (Wells *et al.*, 1990) and west Texas (Gile, 1979). This may indicate a relatively young age for the accumulations and surfaces, or a lack of pedogenic activity as a result of an arid climate.

Development of supersurfaces

Surface A separates the wind-ripple deposits of Unit I from the structureless sands of Unit II. Similar textural characteristics and remnant wind-ripple laminations in Unit II indicate that both units accumulated in a similar manner. In individual trenches Surface A might be taken for a reactivation surface, but its occurrence in similar topographic and stratigraphic positions at widely separated locations indicates that it is of regional extent. In addition, the vesicular A horizon and palaeosol that underlie it indicate that the sand surface was stable for some period of time. Surface A exhibits topographic relief of 1–2 m and follows the present surface. There are no lag deposits associated with Surface A, and primary sedimentary structures in Unit II have been destroyed by pedogenesis. Surface A appears to represent a former land surface developed on a gently undulating sand sheet that was stable and not subject to significant deflation, similar to those found in eastern parts of the sand sea today (Fig. 14). This implies an increase in vegetation cover, which protected the surface and also cut off sand supply. A climatic change to conditions of increased rainfall is implied. Surface A is thus similar to the regional-scale bounding surfaces described by Talbot (1985) and Vökel (1989) that resulted from stabilization of late Pleistocene dunes by vegetation and pedogenesis in the early Holocene period of increased rainfall in the Sahel. It can be classified as a stabilized surface in the terminology of Kocurek & Havholm (in press).

The age of Surface A and the climatic changes that led to its development are unknown at present. Palaeobotanical evidence from the Tinajas Altas Mountains 50 km north of the area (Van Devender, 1990) indicates increased winter rainfall and cooler temperatures coeval with the late Wisconsin 13 000–14 000 BP. However, data from the Yuma area northwest of the Gran Desierto indicate continued dry conditions in the lower Colorado Desert at this time (Cole, 1986). During the mid-Holocene, there

Fig. 14. Vegetation-stabilized sand sheet in eastern Gran Desierto. Such sand sheets are currently forming a stabilized supersurface, similar to that represented by Surface A.

is evidence from a variety of sources for increased summer rainfall and development of ephemeral grasses throughout the lower Sonoran Desert (Van Devender, 1990), and it is possible that Surface A dates to this period.

Surface B is a very extensive planar surface that underlies sand sheet Units I and II as well as the equivalent dune units northwest of the main star dune area. It is characterized by a thin (one to two grain diameters), very coarse sand or granule lag. The underlying Unit III is, however, composed largely of fine and medium sand with a low percentage (<5%) of coarse and very coarse sand. The fine-grained nature of this unit suggests that it is not a 'trailing margin' deposit left behind by the migration of the dunes of the sand sea towards the southeast. The amount of deflation implied by the concentration of coarse grains to form a lag suggests that the surface was exposed for a considerable period of time in the absence of significant vegetation cover. This is supported by the degree of pedogenic alteration, including the development of caliche, in Unit III. Likewise, the absence of significant topography on this surface suggests minimal vegetation cover. Supersurface B therefore appears to have been developed as a result of the termination of sand input to the sand sea due to non-climatic causes, and is a deflationary or erosional supersurface developed in conditions of a negative sediment budget for the area.

The major source for the sands of the western Gran Desierto Sand Sea was probably point bars and river terrace deposits of the lower Colorado River downstream from Yuma (Blount & Lancaster, 1990). Termination of sand supply from this source may have been the result of eustatic changes in sea level that changed the pattern of fluvial sedimentation in the delta system. For example, during the last Glacial Maximum the delta front was located as much as 100 km southeast of its modern position (van Andel, 1964), resulting in the formation of a deeply incised valley in the vicinity of the present delta region. In the absence of a sand input from the Colorado River, 'cannibalization' of existing aeolian deposits then took place with deflation continuing until a surface lag formed. Once the surface had stabilized, pedogenic processes were able to alter pre-existing deposits. It is probable that similar processes are operating today in the northwestern Gran Desierto Sand Sea, now that sand supply from the lower Colorado River valley has been terminated by agricultural development.

SUMMARY AND CONCLUSIONS

Sand sheets in the northwestern part of the Gran Desierto in Mexico have developed as a result of deposition by wind ripples in areas of sparse desert vegetation and an undersaturated sand transport regime (wind energy > sand supply). Changes in climate and/or sand supply gave rise to the development of regional-scale or super bounding surfaces. Termination of sand supply because of eustatic changes but with no coeval climate change led to the development of a supersurface dominated by a granule lag. Climatic changes leading to an increase in the vegetation cover terminated deposition and gave rise to pedogenic alteration of existing deposits, forming a second super bounding surface that was subsequently buried by modern wind-ripple deposits.

These studies indicate the very subtle relationships that exist between sand supply, vegetation cover and aeolian processes in this region. They also demonstrate that supersurfaces formed in different environments can have distinct features. Ongoing investigations will seek to understand the development of supersurfaces in sedimentary environments dominated by dunes, and to reconstruct the history of the sand sea.

ACKNOWLEDGEMENTS

The study was supported by National Science Foundation Grant EAR 89-01457. I thank Gary Kocurek and Steve Fryberger for their constructive reviews and comments.

REFERENCES

ANDEL, T.H. VAN (1964) Recent marine sediments of the Gulf of California. In: *Marine Geology of the Gulf of California* (Ed. Andel, T.H. van & Shor, G.C.) pp. 216–310. American Association of Petroleum Geologists, Tulsa.

ASH, J.E. & WASSON, R.J. (1983) Vegetation and sand mobility in the Australian desert dunefield. *Z. Geomorph. Suppl.* **45**, 7–25.

BLOUNT, G. & LANCASTER, N. (1990) Development of the Gran Desierto Sand Sea. *Geology* **18**, 724–728.

CHURCHWARD, H.M. (1963) Soil studies at Swan Hill, Victoria, Australia. III. Some aspects of soil development on aeolian materials. *Aust. J. Soil Res.* **1**, 117–128.

COLE, K.L. (1986) The Lower Colorado River Valley: a Pleistocene desert. *Quatern. Res.* **25**, 392–400.

COLLETTA, B. & ORTLIEB, L. (1984) Deformations of the middle and late Pleistocene deltaic deposits at the mouth of the Rio Colorado, northwestern Gulf of California. In: *Neotectonics and Sea Level Variations in the Gulf of California Area: a Symposium* (Ed. Malpica-Cruz, V.E.A.) pp. 31–53. University Nacional Auton Mexico, Institute Geolocia, Mexico City.

FRYBERGER, S., AHLBRANDT, T. & ANDREWS, S. (1979) Origin, sedimentary features, and significance of low-angle eolian 'sandsheet' deposits, Great Sand Dunes National Monument and vicinity, Colorado. *J. Sed. Petrol.* **49**, 0733–0746.

FRYBERGER, S.G., AL-SARI, A.M., CLISHAM, T.J., RIZOI, S.A.R. & AL-HINAI, K.G. (1984) Wind sedimentation in the Jafarah sand sea, Saudi Arabia. *Sedimentology* **31**, 413–431.

GILE, L.H. (1979) Holocene soils in eolian sediments of Bailey County, Texas. *Soil Sci. Soc. Am. J.* **43**, 994–1003.

GILE, L.H., PETERSON, F.F. & GROSSMAN, R.B. (1966) Morphological and genetic sequences of carbonate accumulation in desert soils. *Soil Sci.* **101**, 347–360.

HESP, P. (1983) Morphodynamics of the incipient foredunes in New South Wales, Australia. In: *Eolian Sediments and Processes* (Ed. Brookfield, M.E. & Ahlbrandt, T.S.) pp. 325–342. Elsevier, Amsterdam.

KOCUREK, G. (1988) First order and super bounding surfaces in eolian sequences — bounding surfaces revisited. *Sed. Geol.* **56**, 193–206.

KOCUREK, G., DEYNOUX, M., BLAKEY, R.C. & HAVHOLM, K.G. (1990) Amalgamated deposits resulting from climatic and eustatic changes, Akchar Erg, Mauritania. *Abstracts 13th Int. Sed. Congress, Nottingham*, 278.

KOCUREK, G. & HAVHOLM, K.G. (in press) Eolian sequence stratigraphy — a conceptual framework. In: *Recent Advances in Siliciclastic Sequence Stratigraphy* (Ed. Posamentier, H. & Weimer, P.). Am. Assoc. Pet. Geol., Tulsa.

KOCUREK, G., HAVHOLM, K.G., DEYNOUX, M. & BLAKEY, R.C. (1991) Amalgamated accumulations resulting from climatic and eustatic changes, Akchar Erg, Mauritania. *Sedimentology* **38**, 751–772.

KOCUREK, G. & NIELSON, J. (1986) Conditions favourable for the formation of warm-climate aeolian sandsheets. *Sedimentology* **33**, 795–816.

KOCUREK, G., TOWNSLEY, M., YEH, E., HAVHOLM, K. & SWEET, M.L. (1992) Dune and dunefield development on Padre Island, Texas, with implications for interdune deposition and water-table-controlled accumulation. *J. Sed. Petrol.* **62**, 622–635.

LANCASTER, N. (1990) Palaeoclimatic evidence from sand seas. *Palaeogeog., Palaeoclimatol., Palaeoecol.* **76**, 279–290.

LANCASTER, N., GREELEY, R. & CHRISTENSEN, P.R. (1987) Dunes of the Gran Desierto Sand Sea, Sonora, Mexico. *Earth Surf. Proc. Landf.* **12**, 277–288.

NIELSON, J. & KOCUREK, G. (1986) Climbing zibars of the Algodones. *Sed. Geol.* **48**, 1–15.

PORTER, M.L. (1986) Sedimentary record of erg migration. *Geology* **14**, 497–500.

RUBIN, D.M. & IKEDA, H. (1990) Flume experiments on the alignment of transverse, oblique and longitudinal dunes in directionally varying flows. *Sedimentology* **37**, 673–684.

SWEET, M.L., NIELSON, J., HAVHOLM, K. & FARRALLEY, J. (1988) Algodones dune field of southeastern California: case history of a migrating modern dune field. *Sedimentology* **35**, 939–952.

TALBOT, M.R. (1985) Major bounding surfaces in aeolian sandstones: a climatic model. *Sedimentology* **32**, 257–266.

VAN DEVENDER, T.R. (1990) Late Quaternary vegetation and climate of the Sonoran Desert, United States and Mexico. In: *Packrat Middens: the Last 40 000 Years of Bioyic Change* (Ed. Betancourt, J.L., Devender, T.R. van & Martin, P.S.) pp. 134–166. University of Arizona Press, Tucson.

VAN DEVENDER, T.R., THOMPSON, R.S. & BETANCOURT, J.L. (1987) Vegetation history of the deserts of southwestern North America; the nature and timing of the

Late Wisconsin–Holocene transition. In: *North America and Adjacent Oceans During the Last Deglaciation* (Ed. Ruddiman, W.F. & Wright, H.E., Jr) pp. 323–352. Geological Society of America, Boulder.

Vökel, J. (1989) Formation of dunes and pedogenesis as palaeoclimatic indicators in the eastern part of the Republic of Niger (Sahara and Sahel). *Pal. Afr.* **20**, 37–54.

Wells, S.G., McFadden, L.D. & Schultz, J.D. (1990) Eolian landscape evolution and soil formation in the Chaco dune field, southern Colorado Plateau, New Mexico. In: *Soils and Landscape Evolution. Proceedings of the 21st Binghamton Symposium in Geomorphology* (Ed. Knuepfer, P.L.K. & McFadden, L.D.) pp. 517–546. Elsevier, Amsterdam.

Winkler, C.D. & Kidwell, S.M. (1986) Paleocurrent evidence for lateral displacement of the Pliocene Colorado River delta by the San Andreas fault system, southeastern California. *Geology* **14**, 788–791.

Ancient Aeolian Environments

Spec. Publs Int. Ass. Sediment. (1993) **16**, 87–107

Aeolian genetic stratigraphy: an example from the Middle Jurassic Page Sandstone, Colorado Plateau

K.G. HAVHOLM*, R.C. BLAKEY†, M. CAPPS†, L.S. JONES†, D.D. KING† *and* G. KOCUREK*

**Department of Geological Sciences, University of Texas at Austin, Austin, TX 78713, USA; and †Department of Geology, Northern Arizona University, Flagstaff, AZ 86011, USA*

ABSTRACT

Regional tracing and correlation of super bounding surfaces and the aeolian accumulations they cap show that the Middle Jurassic Page Sandstone consists of a three-dimensional mosaic of these basic aeolian building blocks. These basic supersurface bounded units can be grouped into six architectural assemblages termed the Basal, Lower, Isolated, Middle, Upper and Transition Units. The Basal Unit is a complex, discontinuous assemblage formed on the irregular J-2 regional unconformity and may not represent genetically related strata; scouring associated with emplacement of the overlying unit indicates that it was initially more widespread. The main body of the Page consists of three assemblages. The Lower Unit pinches out against the rising J-2 surface along the Monument Bench, and is bounded in a stepwise fashion by supersurfaces that show the progressive eastward transgression of the marine Judd Hollow Member of the Carmel Formation with supersurfaces extending inland. The Isolated Unit, occurring as outliers, is tentatively correlated with the Basal and Lower Units. The Middle Unit intertongues throughout with the Carmel Formation, and consists of the amalgamation of layers and lenses of aeolian units separated by multiple truncational supersurfaces, some of which can be traced into tongues of the Carmel. The Upper Unit of the Page is locally deeply scoured by large troughs, whereas a Transition Unit associated with the onset of Carmel deposition varies from a thin reworked zone to local aeolian build-ups. Throughout the Page, supersurfaces are characterized by polygonal fractures, relief and superposed sabkha and subaqueous strata. Identification of Page genetic units demonstrates a means of internal correlation within aeolian units, and provides a way to link internal aeolian stratigraphy to external events manifested in adjacent non-aeolian units.

INTRODUCTION

That the accumulation of an aeolian succession can be an episodic process was first recognized with Pleistocene–Holocene aeolian successions that clearly represent more than one period of accumulation (e.g. Glennie, 1970; McClure, 1978; Anton, 1983). Surfaces representing hiatal events in aeolian accumulation were first interpreted in an ancient unit, the Permian Cedar Mesa Sandstone, by Loope (1985). Since then examples of aeolian accumulations separated by surfaces have been recognized in a variety of units ranging in age from the Proterozoic to the Holocene (e.g. Talbot, 1985; Fryberger & Koelmel, 1986; Deynoux *et al.*, 1989; Kocurek *et al.*, 1991a). Talbot (1985) and Kocurek (1984, 1988) suggested that these surfaces, termed super bounding surfaces, should be useful as a correlation tool within aeolian units that had previously been considered as lacking any means of internal correlation.

Kocurek & Havholm (1992) refine the definition of a supersurface to represent the end of an aeolian accumulation event. They believe that the hiatus in aeolian accumulation represented by a supersurface

* Present address of first author: Geological Sciences, AJ-20, University of Washington, Seattle, WA 98195, USA.

is caused by a change in the sediment budget conditions in the aeolian environment and therefore a single accumulation and its capping supersurface represent the basic unit or building block from which an aeolian succession is built. This basic aeolian unit has elements of similarity to other types of stratigraphic unit. Like the 'genetic stratigraphic sequence' of Galloway (1989), each basic aeolian unit represents a distinct, coherent episode of sedimentation, a genetic unit, and it is with this sense that the term 'genetic' is used throughout this study. Like the 'sequence' of sequence stratigraphers (Van Wagoner *et al.*, 1988), each basic aeolian unit is defined by bounding surfaces formed during subaerial exposure. However, in scale the basic genetic aeolian unit may be more comparable to parasequences or parasequence sets than to the larger sequences. In addition, the factors that control coastal/marine sequence or parasequence formation are not exactly the same as those that control aeolian stratigraphic patterns (Havholm, 1991; Kocurek & Havholm, 1993).

The purpose of this study is to demonstrate that supersurfaces can be used on a regional scale to help separate an aeolian formation into its genetically distinct parts, that is, packages of rock that represent periods of accumulation and the supersurfaces that separate them and represent periods of hiatus in accumulation. This puts aeolian stratigraphy on an event–stratigraphic footing, allowing recognition of the events associated with the development of an aeolian succession as well as making future integration with non-aeolian sequence stratigraphy more feasible. Future studies will deal with topics such as the distribution and significance of the features on the surfaces, interpretation of the events represented in the Page Sandstone, and the causes of and controls on the Page events, and these issues are not addressed here. Hence, what follows is a case-study in which the three-dimensional genetic–stratigraphic architecture of the Middle Jurassic (Bajocian–Bathonian) Page Sandstone is defined on the basis of supersurface geometry, the stratification style of the accumulations bounded by the supersurfaces and the relationships of these units to one another and adjacent non-aeolian units.

PAGE SANDSTONE

The Page Sandstone has been interpreted as a coastal erg that lay along the Carmel inland seaway near the western edge of the Middle Jurassic North American craton, lapping onto the Monument Upwarp (Fig. 1), an area that was a structural bench during early Middle Jurassic time (Peterson, 1986). The Page Sandstone was chosen for this study for several reasons. It has large expanses of excellent, generally accessible exposures. It covers a relatively small geographical area by Colorado Plateau standards and is nowhere thicker than 100 m (Fig. 2). It contains multiple laterally extensive surfaces that have features that generally distinguish them from other relatively featureless, less extensive surfaces in the Page Sandstone. These surfaces have been interpreted as supersurfaces based on their lateral continuity and their distinctive features and because they generally separate aeolian strata representing dunes of different sizes and types (Havholm, 1991; Kocurek *et al.*, 1991b). The surface features include pronounced local relief associated with deflation of a damp or cemented surface, the presence of overlying wavy-laminated, fine-grained sabkha deposits and abundant fractures that in plan view are polygonal. These polygonal fractures have been interpreted as the result of expansion and contraction of an evaporite-cemented substrate (Kocurek & Hunter, 1986). Moreover, the Page Sandstone inter-fingers with and is overlain by the marine and non-aeolian coastal plain sediments of the Carmel Formation, allowing the potential of correlating aeolian accumulations and their capping surfaces with their equivalent expression in a non-aeolian unit.

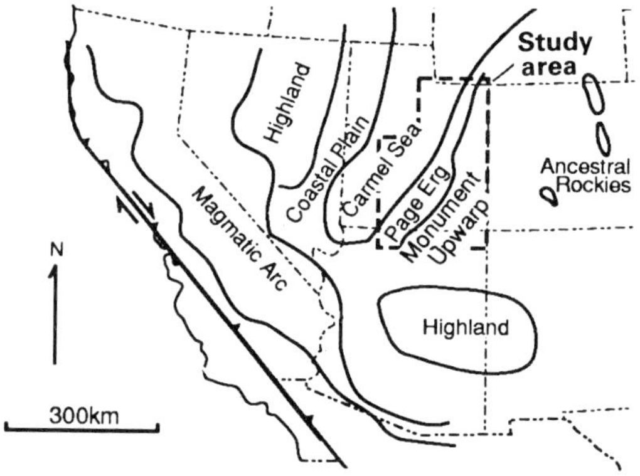

Fig. 1. General palaeogeographic setting of the Middle Jurassic Page Erg and area of this study, modified from Blakey *et al.* (1983) and Saleeby & Busby-Spera (1992).

Considerable work providing the lithofacies description, environmental interpretation and tectonic and stratigraphic framework of the Page and Carmel Formations can be found in Voorhees (1978), Geesaman (1979), Peterson & Pipiringos (1979),

Caputo (1980), Blakey *et al.* (1983), Peterson (1986), Blakey *et al.* (1988) and Kocurek *et al.* (1991b). The detailed work from which much of this paper is summarized is presented in Capps (1990), Jones (1990), Havholm (1991) and King (1992).

In the course of this project, 321 sections were measured (Fig. 2; Appendix Table A1). Figure 3A (facing p. 88) illustrates the three-dimensional architecture of the Page Sandstone by way of a fence diagram whose location is shown on Fig. 2A. Some more detailed traverses are shown on Fig. 4A–E (Fig. 4D & E face p. 112); their locations, which differ slightly from the traverses shown in the fence diagram in Fig. 3A, are indicated on Fig. 2C. In some cases, surfaces and units had to be correlated between sections, but in other cases they were traced either by physically walking the surfaces or by following them on photographic panoramas.

The stratigraphic framework defined by the genetic architecture of the Page Sandstone is at odds with the formal stratigraphic divisions of the Page Sandstone. A comparative chart (Fig. 3B) shows how the major supersurface bounded units, described below, relate to the formal stratigraphic divisions. Formally, in the west and south the Page Sandstone is subdivided into two members by the Judd Hollow Tongue of the Carmel Formation; the part of the Page Sandstone underlying the Judd Hollow is the Harris Wash Member and that overlying it is the Thousand Pockets Member. Elsewhere the Page Sandstone is undivided; however, Peterson & Pipiringos (1979) recognized that the Page Sandstone found along the Waterpocket Monocline (southeast of Teasdale, sections TSD–HCB, Fig. 2A) and in the Henry Mountains and San Rafael Swell (east and north of Teasdale, Fig. 2A) was equivalent to the Harris Wash.

GENETIC STRATIGRAPHIC UNITS AND SUPERSURFACES

The Page Sandstone everywhere rests on the J-2 unconformity, one of six regional unconformities recognized by Peterson & Pipiringos (1979). The Page Sandstone is thickest at its southern, erosionally truncated edge, and generally thins northeastward, and northwestward along its seaward margin where it inter-fingers with and is replaced by the Carmel Formation (Figs 2 & 3A & B). Along its southeastern, inland margin it thins by erosion and by onlap (sections MNW–TSG, Fig. 4B) onto the Monument Upwarp (Fig. 1), a structural bench during the Jurassic (Peterson, 1986).

The Page Sandstone is here subdivided into six major units, some of which are composed of smaller genetic units. Figure 5A–C shows an example of each of these units. The oldest Page strata above the J-2 surface comprise a distinctive, discontinuous unit referred to here as the Basal Unit. The younger main body of the Page Sandstone consists of three primary architectural units: Lower, Middle and Upper Units. A unit found only in outliers of the Page Sandstone (Isolated Unit) is tentatively correlated with the lower part of the Page Sandstone, including part or all of the Basal, Lower and lower Middle Units. The Upper Unit has filled scours cut into its upper surface. The Page Sandstone is capped in many areas by a Transition Unit associated with the onset of Carmel accumulation across much of the region. The names of the six major units of the Page Sandstone are capitalized here for ease of communicating our ideas and we do not mean to imply that these should be taken as formal stratigraphic subdivisions.

Basal Unit (PBd, PBs), basal surface (S--b)

The lowermost genetic accumulation of the Page Sandstone (PBd, PBs) occurs as a thin unit (5 m thick or less except in the Page, Arizona area where it is up to 15 m thick), with two distinct lithofacies (PBd, PBs). In the south, the Basal Unit (PBd) is a typically bimodal, fine- to medium-grained sandstone showing wavy laminae, low-angle wind-ripple strata and less common intervals with adhesion-ripple structures and sets of high-angle dune cross-strata. These features alternate vertically and grade laterally into one another, representing a patchwork accumulation of aeolian sand sheets, sabkhas and isolated dunes (Havholm, 1991). In some areas, surfaces with polygonal fractures and/or highly corrugated relief occur within this unit. In the north (Hanksville–Ticaboo, sections HHS–SFT, and PTK, Fig. 3A), what is believed to be the equivalent Basal Unit is represented by a clayey–sandy siltstone with sandstone layers and lenses (PBs). The siltstone locally contains siltstone rip-up clasts and halite crystal moulds; the sandstone shows wavy laminae and breccias of wind-ripple strata. This lithofacies is believed to have been deposited in a muddy sabkha to a possible fluvial and evaporative lacustrine environment.

The Basal Unit is complex and not necessarily a single genetic unit. It is typically capped by a

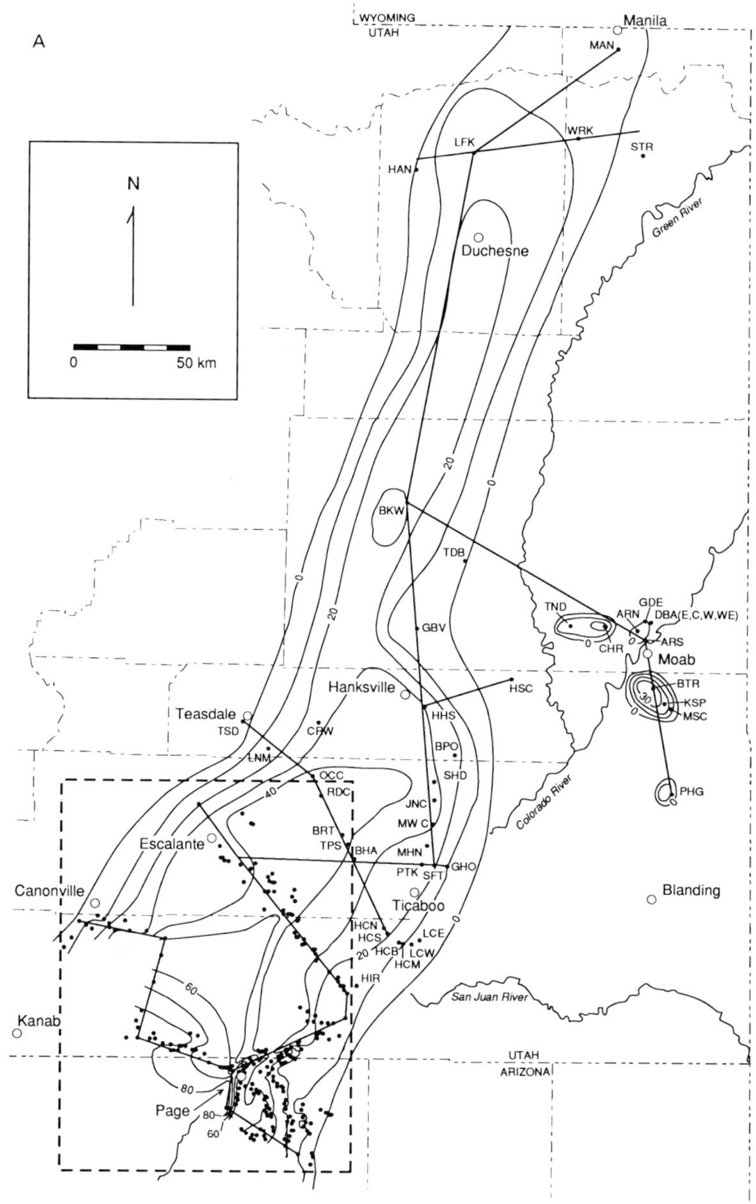

Fig. 2. (A) Isopach map of Page Sandstone; contours show total thickness of the Page Sandstone in metres (including the thickness of tongues of the Carmel Formation that are within the Page Sandstone). Dots are locations of measured sections; three-letter labels are abbreviations for names of measured sections (see Appendix for section names and locations). Traverse lines act as the base of the fence diagram in Figs 3A & B. The area within the dashed rectangle is enlarged in Fig. 2B. (B) Enlargement of southern end of Page Sandstone outcrop area showing section abbreviations (see Appendix) and isopach contours (in metres). (C) Map of the southern end of the Page Sandstone outcrop area showing the locations of the traverses illustrated in Figs 4A–E.

supersurface (S--b) with well developed polygonal fractures. As seen on Fig. 4B, D & E, in some cases the Basal Unit is confined to topographic depressions on the J-2 surface (unit A in Kocurek *et al.*, 1991b), and this may represent the original depositional extent of the unit. Elsewhere, however, the unit is clearly truncated by erosion associated with troughs scoured during accumulation of the overlying Lower Unit, and here it may have originally been more extensive (Fig. 4B, D & E). The silty facies (PBs) is more continuous than the unit in the south, but it too pinches out against palaeotopographic high areas on the J-2 surface (e.g. northwest of section SHD, Fig. 3A). As discussed in Blakey *et al.* (1988) and Kocurek *et al.* (1991b), facies grouped here as the Basal Unit developed as the initial deposits on the regional J-2 unconformity, and these thin accumulations may have formed at various times

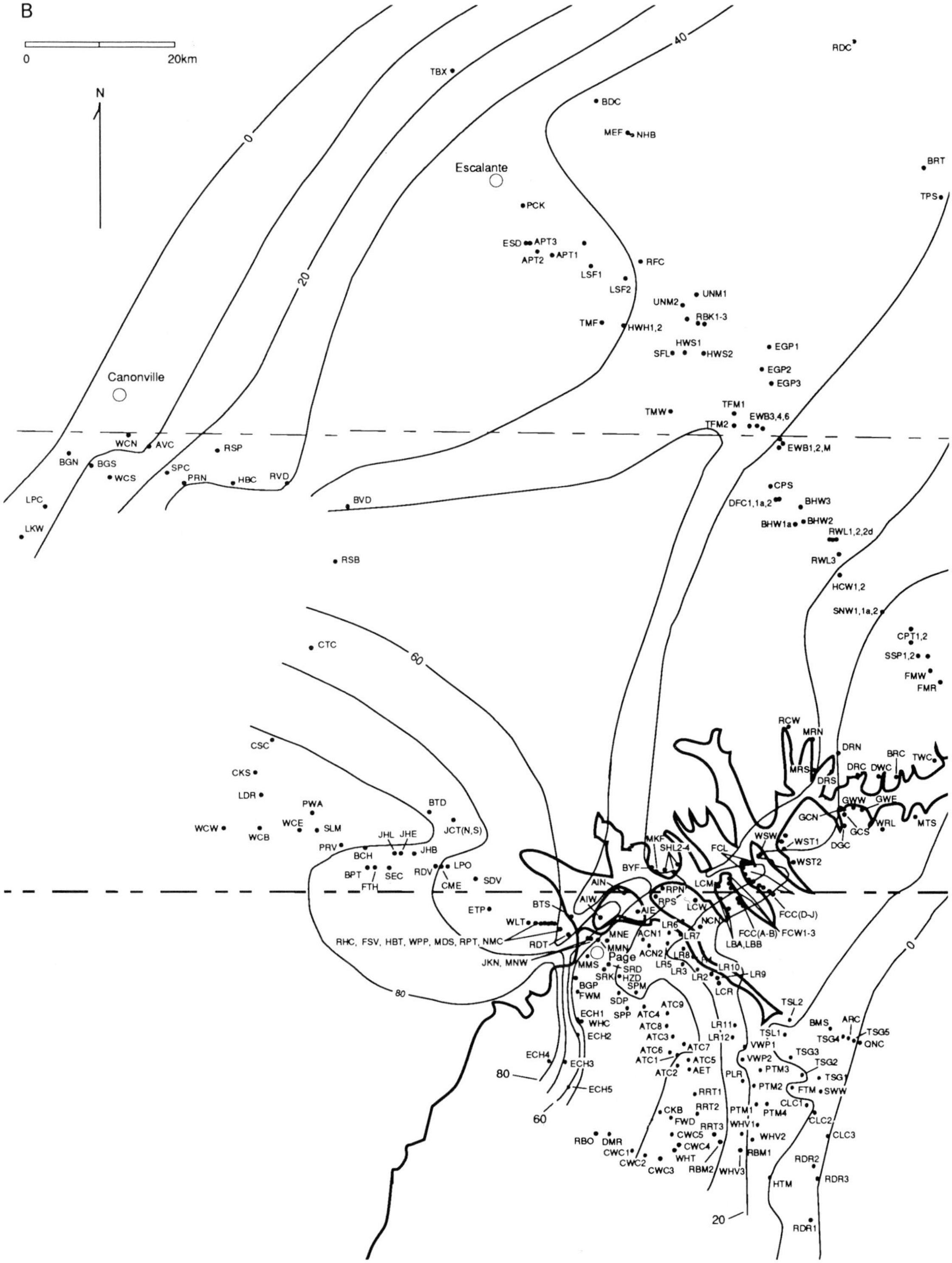

Fig. 2. (continued)

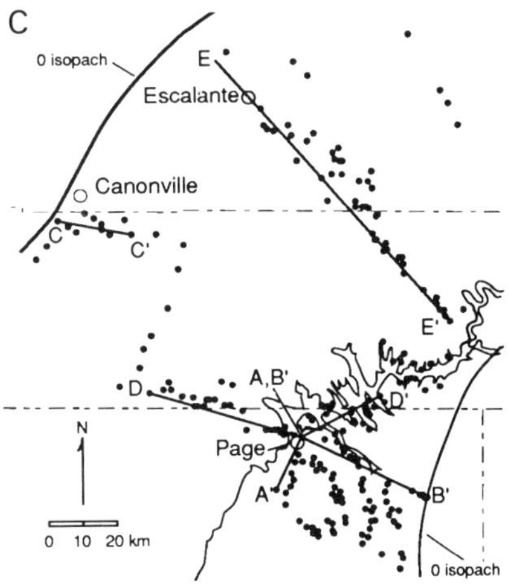

Fig. 2. (continued)

during exposure of the J-2 surface and may represent more than one accumulation event.

Lower Unit (PL, PLf), capping supersurfaces (including S--jh), Erg Margin Unit (PE)

Overlying the Basal Unit (PBd, PBs), or the J-2 unconformity where the Basal Unit is absent, is the Lower Unit (PL, PLf), which forms the geographically most extensive unit of the Page Sandstone (Figs 3A, 4A–E & 6). This unit is thickest in the Harris Wash to Brimhall Arch areas (sections HWS, southeast of Escalante, and BHA, west of Ticaboo, Fig. 3A) and thins and pinches out against the J-2 unconformity along the palaeotopographic high of the Monument Bench along its inland, southeastern margin (along Lake Powell in Figs 1 & 3A, to the southeast in Fig. 4B). Basinward, the Lower Unit is in some areas partly replaced laterally by the lower Middle Unit (PMl) and the Judd Hollow Tongue of the Carmel Formation (CJ) in a stepwise manner best illustrated in Fig. 4E; in other areas the transgressing marine and sabkha units of the Judd Hollow simply overlie the Lower Unit (e.g. the westernmost part, Fig. 4D). To the north along the axis of the Page Sandstone, the Lower Unit thins to the north of Hanksville, but then thickens again in the San Rafael Swell (section BKW, Fig. 3A). Where it is next seen, north of Duchesne, it is thinner again (section LFK, Fig. 3A) and pinches out northward.

The main part of the Lower Unit is a fine- to medium-grained sandstone (PL, Figs 3A & 4A, B, D & E) and consists mainly of cross-strata that represent large sinuous crescentic dunes and draas (Havholm, 1991; units B and C, Kocurek *et al.*, 1991b). In the Cannonville area the Lower Unit is very thin overall and consists of thin, simple aeolian sets and contorted to structureless sandstone (PLf on Figs 3A & 4C). In the Ticaboo area and in the area north of Duchesne, the Lower Unit is replaced laterally by the Erg Margin Unit (PE, Fig. 3A) that consists of a stack of dune cross-strata and sabkha beds, truncated by up to nine supersurfaces; the Erg Margin Unit (PE) is probably also partly equivalent to the lower Middle Unit.

The Lower Unit has a different capping supersurface in different areas. In the west, near Cannonville, it is the surface at the base of the Judd Hollow Tongue of the Carmel Formation (e.g. sections BGN–BVD, Figs 3A & 4C). In the south and southeast it is capped by the JH surface that correlates to the maximum transgression of the Judd Hollow (S--jh, e.g. west of Page, Arizona, sections SDV–RDT, Fig. 3A, east from section WLT, Fig. 4D). This surface laps onto the J-2 unconformity to the southeast of Page, Arizona (Fig. 4B) and is removed by erosion associated with aeolian scouring in the Middle Unit north and east of Wetherill Canyon (section WLC, Fig. 3A). Elsewhere the Lower Unit is capped by one of a series of surfaces extending inland from the advancing transgression that form the base of the Middle Unit (e.g. near Judd Hollow, sections BPT–CME, Fig. 4D; from Escalante to southeast of Early Weed Bench, sections TBX–DFC, Fig. 4E) or by a later supersurface (S--up) that truncates down to the Lower Unit east along Lake Powell (e.g. southeast of Hurricane Wash, sections HWA–FMR, Fig. 4E).

Locally there is a supersurface or group of surfaces (lower surface zone, S--l) within this unit (Figs 3A & 4A, D & E) suggesting that it represents two accumulation events. This surface, or an equivalent surface zone, is discontinuous, truncated by troughs in the overlying unit (Fig. 4D & E). The occurrence of this surface, or the equivalent zone of several surfaces, in widely spaced areas suggests that it may have been a much more widespread surface or surface zone before truncation. The expression of this supersurface varies from area to area. Southeast of Escalante from Early Weed Bench to Fiftymile Ridge (sections EWB–FMR, Figs 3A & 4E) surfaces in this group typically have large polygonal fractures and, more rarely, are overlain by thin sabkha beds. Northwest of Page, Arizona in the Judd Hollow area (WCB–CME, Fig. 3A; BCH–CME, Fig. 4D) it consists of a discontinuous zone of thin sets of aeolian cross-strata separated by sabkha beds and corrugated to polygonally fractured surfaces.

Isolated Unit (PI)

Around Moab are pods of Page Sandstone that are isolated in topographic lows on the J-2 surface (PI, Fig. 3A). These have a particularly low proportion of dry aeolian strata, and include subaqueous ripple strata as well as sabkha sediments (see also Swezey, 1991). Because of their isolation, these localized lenses of Page Sandstone cannot be correlated with confidence to units in the main body of the Page Sandstone. However, based on their sedimentation style, which bears similarity to both the Basal Unit (PBd) and the Erg Margin Unit (PE), and their location far to the north of areas that contain the Upper Unit, the Isolated Unit is probably correlable to part or all of the Basal, Lower and lower Middle Units.

Middle Unit (PMl, PMm, PMu), middle surface zone (S--jh, S--rm, S--ue)

The Middle Unit of the Page Sandstone (PMl, PMm, PMu) inter-fingers with or grades laterally westward into two members of the Carmel Formation to the west, the Judd Hollow Tongue (Jcj, CJ) and the Crystal Creek Member (Jcc). Also included in this part of the Page Sandstone is a marine to sabkha unit ('red marker' of Blakey *et al.*, 1983; CR, this study) that was previously considered to be a tongue of the stratigraphically higher Crystal Creek Member of the Carmel, before Jones (1990) demonstrated that it is an upper tongue of the Judd Hollow Member of the Carmel (Figs 3A & 4C).

The Middle Unit of the Page Sandstone consists of dune strata truncated by up to nine supersurfaces. The resulting package is an amalgamation of layers and lenses of aeolian strata (PMl, PMm, PMu, Fig. 4A–E; fig. 3, Kocurek *et al.*, 1991b; fig. 20a, Havholm, 1991). This unit is thickest to the southwest in the West Clark Bench area (e.g. sections WCB–CSC, Fig. 3A, section PWA, Fig. 4D), where it begins to inter-finger with the Carmel Formation. It thins westward (basinward) as it is replaced by the Judd Hollow Tongue (CJ, CR, Jcj) and the Crystal Creek Member (Jcc) of the Carmel Formation (Figs 3A & 4C & E). To the east and southeast the Middle Unit thins as it is erosionally truncated from above by the S--ue surface (Figs 3A & 4B & D) and by the S--up surface (east end of Fig. 4D & E).

The dune strata in the Middle Unit (PMl, PMm, PMu) represent accumulation by a variety of dune types and sizes, but the dunes were generally smaller than those that produced the underlying Lower Unit (Havholm, 1991; Kocurek *et al.*, 1991b). In various areas of this zone, particularly along the margins of the Page Sandstone, aeolian strata are contorted or slumped, and the proportion of sabkha to aeolian strata also increases towards the margins of the formation. In the Cannonville area rotated mounds and ridges of the Middle Unit mark the edges of the Page Sandstone (Figs 3A & 4C).

The surfaces in the middle zone are highly irregular on a fine scale and truncate one another (e.g. in the Ferry Swale area, sections WLT–NMC, Fig. 4D; see also fig. 3, Kocurek *et al.*, 1991b; fig. 20a, Havholm, 1991). These surfaces have the whole range of features associated with supersurfaces in the Page Sandstone, including polygonal fractures, relief at the centimetre to metre scales and superposed sabkha and, locally, subaqueous sediments.

The Middle Unit is divided into three parts where it inter-fingers with the two tongues of the Judd Hollow (CJ and CR, Figs 3A & 4D & E). In addition, surfaces equivalent to these Carmel Formation tongues can be traced well into the part of the Page Sandstone that is entirely aeolian, subdividing it even in areas the tongues of the Carmel Formation did not reach or where they are not preserved.

Lower Middle Unit (PMl), JH surface (S--jh)

The lower part of the Middle Unit (PMl) underlies the Judd Hollow Tongue of the Carmel (CJ) or the equivalent supersurface (S--jh, Figs 3A & 4D & E). From the Escalante area to the southeast (sections PCK–RWL, Figs 3A & 4E) this unit contains a series of supersurfaces that extend from the advancing Judd Hollow Tongue inland into the erg with stratigraphically higher surfaces extending progressively further inland. Further south in the Judd Hollow area (sections BPT–CME, Fig. 4D), this architectural element is also present but is much less extensive.

The lower Middle Unit (PMl) is probably time-equivalent to the lowermost part of the Judd Hollow Tongue basinward and to the uppermost part of the Lower Unit inland. As the transgression progressed inland, aeolian accumulation of the Lower Unit continued deep in the erg, while at the erg margins

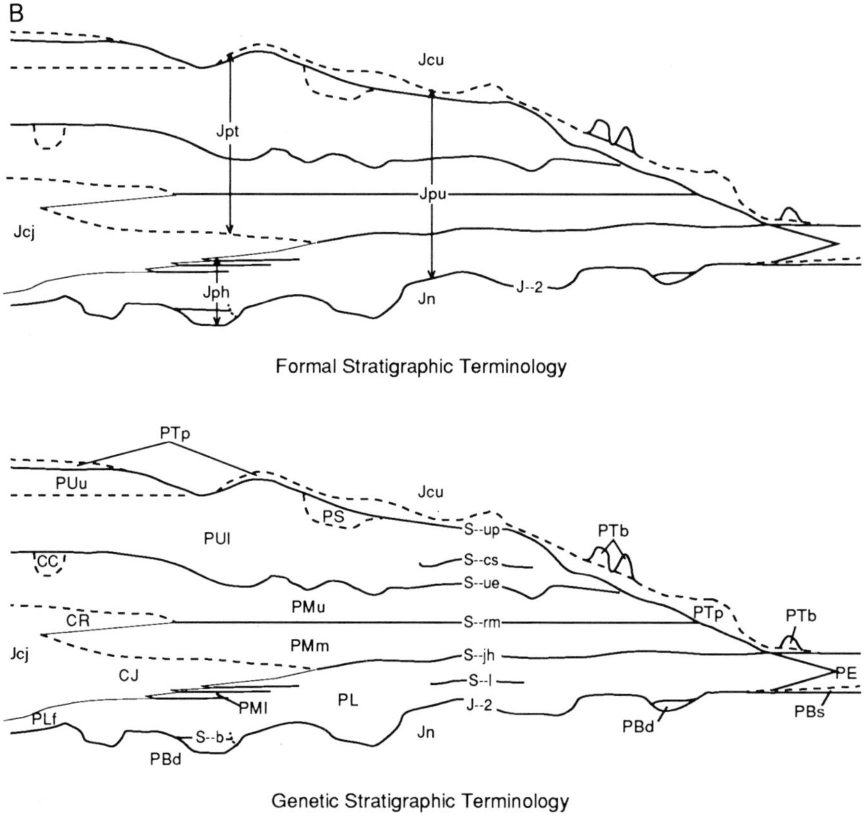

Explanation of Symbols in Figures 3,4,6

Page Genetic Units

PBd	Basal Unit, sandy facies
PBs	Basal Unit, silty facies
PL	Lower Unit, main part
PLf	Lower Unit, marginal flat beds
PM	Middle Unit
PMl	Middle Unit, lower part
PMm	Middle Unit, middle part
PMu	Middle Unit, upper part
PUl	Upper Unit, lower part
PUu	Upper Unit, upper part
PS	Scours
PE	Erg Margin Unit
PI	Isolated Unit
PTp	Transitional Unit, main part
PTb	Transitional Unit, build-ups

Page Super Surfaces

S--b	top of Basal Unit
S--l	within the Lower Unit
S--jh	top of Lower Unit
S--rm	top of middle Middle Unit
S--ue	base of the Upper Unit
S--cs	within the Upper Unit
S--up	top of Upper Unit
J--2	unconformity, base Page

Page Formal Units

Jph	Harris Wash Member
Jpt	Thousand Pockets Member
Jpu	Page, undivided

Carmel Units (this report)

CJ	lower tongue of the Judd Hollow
CR	upper tongue of the Judd Hollow
CC	channel facies cut into Jmu

Carmel Formal Units

Jcj	Judd Hollow Tongue
Jcc	Crystal Creek Member
Jcu	Upper Member
Jn	Navajo Sandstone

Fig. 3B. Diagram illustrating the relationship of the formal stratigraphic units to most of the stratigraphic units as defined in this study. Below is the explanation for all unit abbreviations used in this study.

aeolian accumulation was punctuated by periods of supersurface formation as the marine and sabkha environments approached.

Middle Middle Unit (PMm), RM surface (S--rm)

Between the two tongues of the Carmel (Judd Hollow, CJ, and red marker, CR; Figs 3A & 4C, D & E) is the middle Middle Unit (PMm), an aeolian unit that is bordered on the north and west by sandy and muddy sabkha strata. It is restricted to the southern part of the Page Sandstone, in contrast to the Lower Unit (PL) (Fig. 6). The extent of the middle Middle Unit does not represent its original depositional extent because it is erosionally truncated on its southeastern margin (Figs 3A & 4E), and perhaps also its northeastern margin (Fig. 6), by the upper erosional surface of the Page Sandstone (S--up). In the Lake Powell area, from the Sand Hills east (from section SHL, Fig. 4D), there are troughs scoured through surfaces within this unit. They are up to 12 m deep and from 10 m to more than 100 m across, and many are elongate in a northeast–southwest direction. They occur separately or in clusters with younger ones truncating older ones on the southeast side. These are interpreted as superscoops (Blakey *et al.*, 1988; Blakey & Parnell, in press). These scours (Fig. 4D) are filled with complex irregular to trough cross-stratified aeolian sandstone, and many contain pods of bentonite. On the western margin of the Page Sandstone, near Cannonville, the middle Middle Unit is only locally preserved as isolated mounds of brecciated to distorted to indistinctly laminated sandstone, encased in Judd Hollow strata (near section HBC, Fig. 3A, section RSP, Fig. 4C).

The RM surface (S--rm) extends across the Page Sandstone (Figs 3A & 4A–E) except where it is truncated by scours or by the upper erosional surface of the Page Sandstone (S--up, at Gregory Bay, section GBS, Fig. 4D) and where it laps out onto the J-2 surface (Fig. 4B). The RM surface underlies the red marker (CR), which represents the last major transgression of the Carmel sea into the Page sand sea. This surface continues inland, where it represents a hiatus in aeolian accumulation within the erg that correlates to the transgression. The RM surface is planar to polygonally fractured and commonly overlain by sabkha sediment.

Upper Middle Unit (PMu), UE surface (S--ue)

The margins of the upper Middle Unit (PMu), which lies above the upper tongue of the Judd Hollow and the RM supersurface (S--rm, Figs 3A & 4A–E), are shifted even further south than the middle Middle Unit (PMm), and also to the west (Fig. 6). The upper Middle Unit (PMu) thins to the south, where it is replaced by the Upper Unit (section RDT–ECH, Fig. 3A, MNW–LCK, Fig. 4B). It laps up onto the J-2 unconformity to the southeast (Fig. 4B), but is truncated by the upper erosional surface of the Page Sandstone (S--up) to the northeast of Page, Arizona along Lake Powell (section GBN, Fig. 4D) and southeast of Escalante (section SFL, Fig. 4E). To the west of Page, Arizona the upper Middle Unit thickens (sections WLT–CSC, Fig. 3A; sections SDV–CKS, Fig. 4D). It is not clear whether this eastward thinning is because of truncation (by the S--ue surface) or whether it is replaced laterally by the Upper Unit (sections SDV–CKS, Fig. 4D). Near the erg margin, this upper Middle Unit grades over several kilometres into the Crystal Creek Member of the Carmel Formation (Jcc) to the northwest in the Cannonville area (Figs 3A & 4C). The Crystal Creek Member in this area has been reinterpreted as arid fluvial/alluvial strata (Jones, 1990). In addition, channels that cut into the upper Middle Unit in the Paria area (CC, Figs 3A & 4D) have been reinterpreted to be wadi channels with palaeocurrent indicators showing flow to the northeast (Jones, 1990). The upper Middle Unit (PMu) is very similar to the middle Middle Unit (PMm) in that it consists of erg layers and lenses truncated by supersurfaces, has superscoops within it, and its westernmost expression is in isolated, rotated blocks of aeolian and sabkha strata partly encased by Judd Hollow strata (section BGS, Fig. 3A; section WCN, Fig. 4C).

Truncating the upper portion of the Middle Unit (PMm, PMu), at least in the southern part of the Page Sandstone outcrop area, is the UE surface (S--ue), overlying which is the Upper Unit of the Page Sandstone (Figs 3A & 4A–D). Although this surface does not generally have polygonal fractures, it is in many areas coincident for some distance with various polygonally fractured surfaces of the Middle Unit. The UE surface (S--ue) has irregular relief (e.g. along Lake Powell from the Sand Hills to Face Canyon, sections SHL4–FBS, Fig. 4D), and to the east and south the UE surface is truncated by the UP surface (S--up, e.g. at Face Canyon, section FBN2, Fig. 4D, and near Tse Esgizzi, sections BMS1–BMS2, Fig. 4B).

Upper Unit (PUl, PUu), scours (PS) and upper supersurfaces (S--cs, S--up)

The Upper Unit of the Page Sandstone (PUl, PUu) is only present to the south (Fig. 6), and is thickest around West Clark Bench (WCB, CKS), Page, Arizona (RDT, MNW) and to the south of Page, Arizona (ECH, Figs 3A & 4A & D). It thins by erosional truncation to the southeast and northeast of Page, Arizona (Figs 3A & 4B & D). To the west of Page, Arizona it is locally thinned by erosion from the top and by gradual replacement of its base by the upper Middle Unit (Figs 3A & 4D).

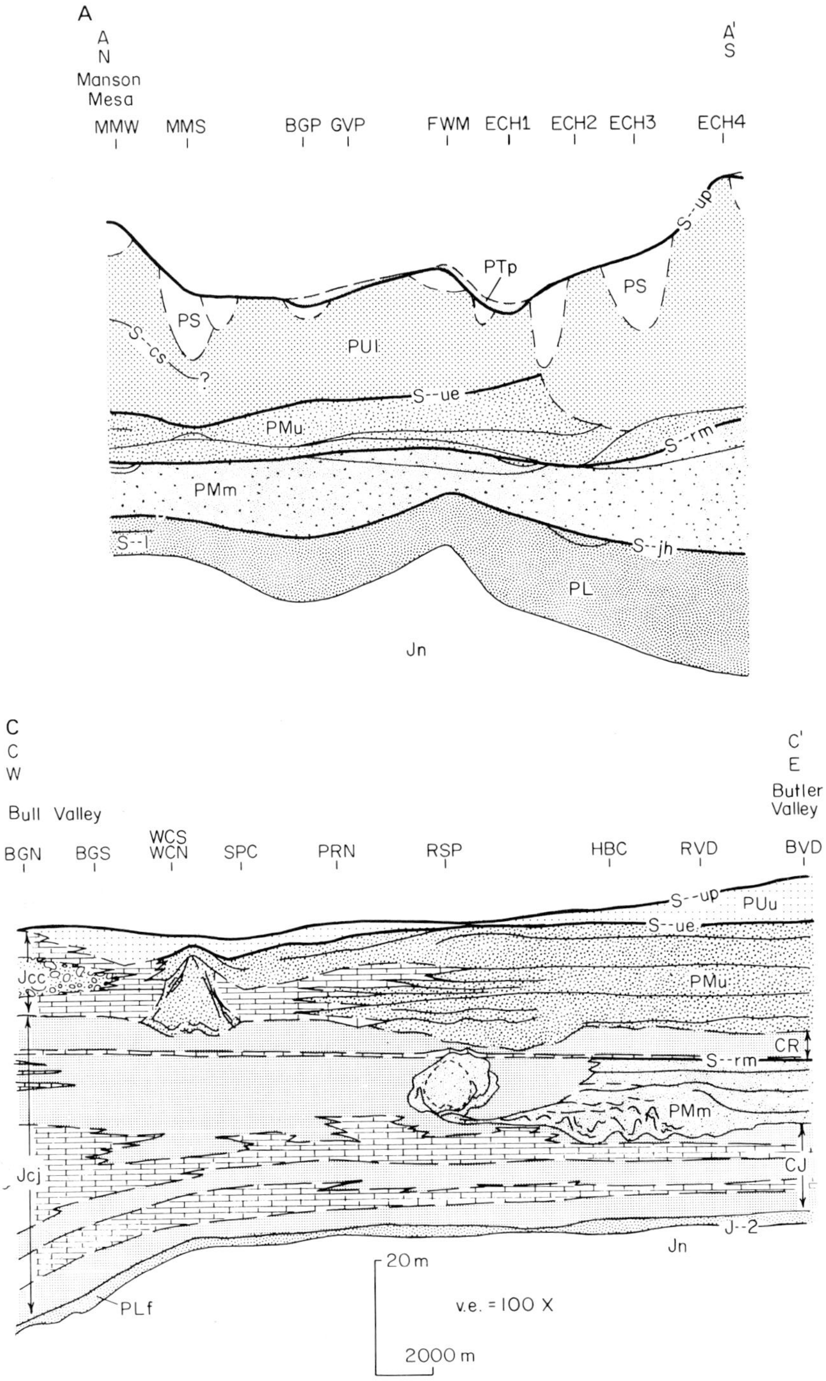

Fig. 4. (A)–(E) Cross-sections of Page Sandstone; location map is Fig. 2C. Fig. 4D & E face p. 112. Solid lines represent supersurfaces; dashed lines are other surfaces. Bold solid lines are the surfaces separating the major Page Sandstone units as described in the text. In the Carmel Formation units, carbonates are shown with a brick pattern, siltstones and silty sandstones are shown with a dotted pattern, and conglomerates with a gravel pattern. Page units consist mainly of aeolian sandstone, but locally also include contorted to structureless sandstones.

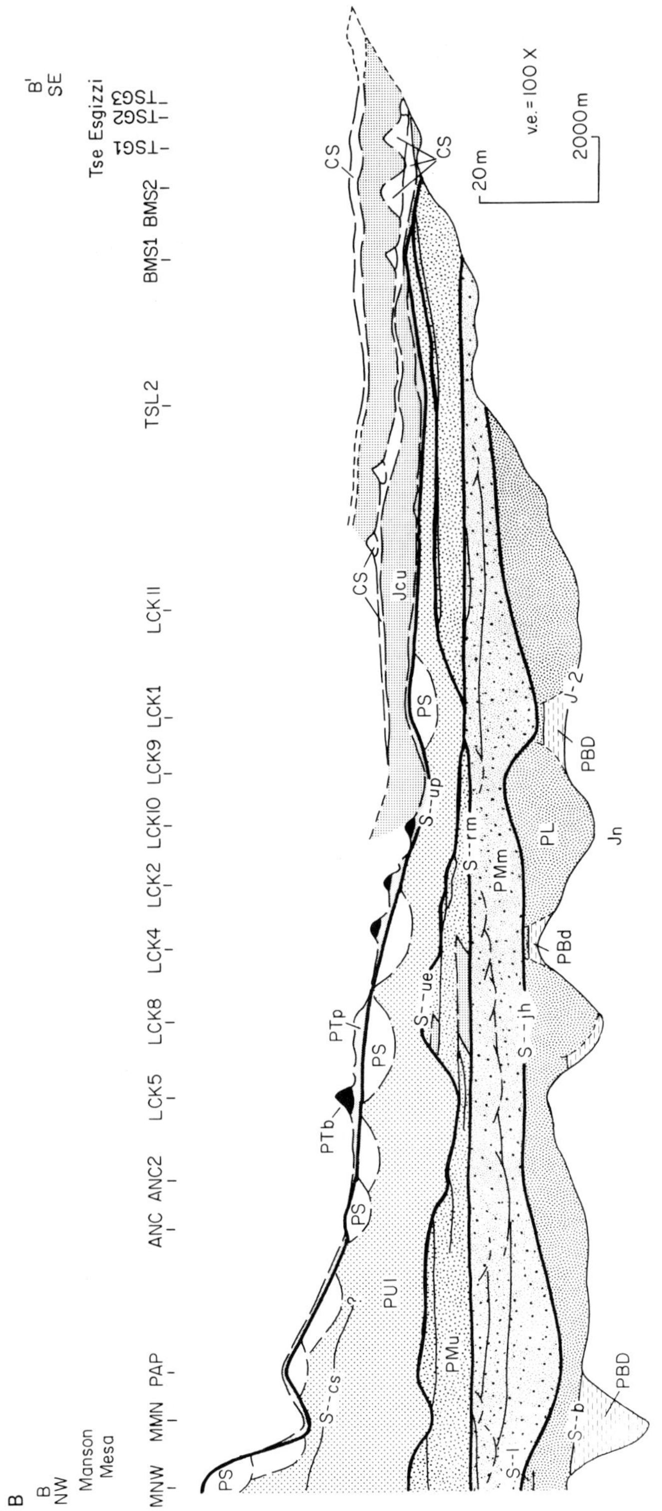

Fig. 4. (continued)

A

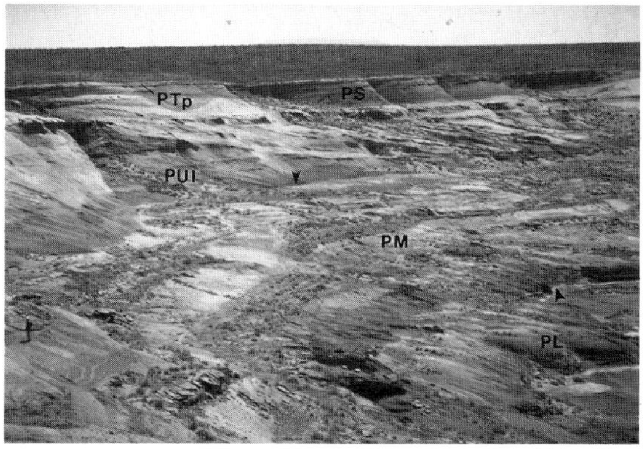

B

C

Fig. 5. (A) The lower three-quarters of the Page Sandstone at Ferry Swale (west from section FSV, Fig. 4D). Arrows point to, from bottom to top, the J-2 unconformity, the S--b surface, the S--jh surface and the S--rm surface. See Fig. 3B for explanation of rock unit symbols. Thickness of the Basal (PBd) and Lower (PL) Units together here is approximately 10 m. (B) Upper three-quarters of the Page Sandstone at Ferry Swale (east from section WPP, Fig. 4D). Lower arrow points to the S--jh surface; upper arrow points to the S--rm surface. See Fig. 3B for explanation of rock unit symbols. Note person in lower left for scale. (C) Isolated Unit (PI) of the Page Sandstone at Double Arch (section DBA) in Arches National Park. Lower arrow points to the J-2 unconformity; the upper arrow points to the Page–Carmel contact.

This Upper Unit has two main lithofacies. The lower one (PUl) comprises all of the Upper Unit except from west of the Judd Hollow area (JHL) to the Cannonville area, where it is overlain by the upper Upper Unit (PUu, west of section JHL, Figs 3A & 4D). The lower Upper Unit (PUl) is replaced from the base by the upper Middle Unit (PMu) northwestward until only the upper Upper Unit (PUu) remains in the area around Cannonville (Figs 3A & 4C & D).

The lower Upper Unit (PUl) consists of thick sets (up to 25 m) of simple and compound cross-strata representing very large draas (Havholm, 1991) and is finer-grained than the underlying units (very fine to fine sand). In the area just east of Page, Arizona, a surface has been recognized within the lower Upper Unit (PUl) that at least locally divides it into two parts (S--cs, Fig. 4A, B & D). This surface has 9–15 m of relief on it from Antelope Island (AIN) eastward (Fig. 4D). The upper Upper Unit (PUu) consists of thinner sets of simple dune cross-strata with planar bounding surfaces, locally interspersed with sabkha strata and/or truncated by corrugated surfaces.

From just west of Page, Arizona, east to Antelope Island (sections WPP–AIN, Fig. 4D) and to the south (Fig. 4A & B), trough-shaped scours (PS) cut into the Upper Unit. These scours are up to 30 m deep and 3 km wide and may occur in bundles en-echelon with younger scours to the southeast. They are mostly filled with aeolian strata but locally contain sabkha strata and bentonite layers. These scours are interpreted as superscoops (Blakey, 1988; Blakey & Parnell, in press).

An erosional supersurface (S--up) caps the Upper Unit (PUl, PUu) and the filled scours (PS) in the south and southwest part of the Page Sandstone outcrop area. This UP surface truncates the Upper Unit entirely northeast (section FBS, Fig. 4D) and southeast (section BMS1, 2, Fig. 4B) of Page, Arizona and cuts down into the upper Middle Unit, also truncating it entirely at Gregory Bay (section

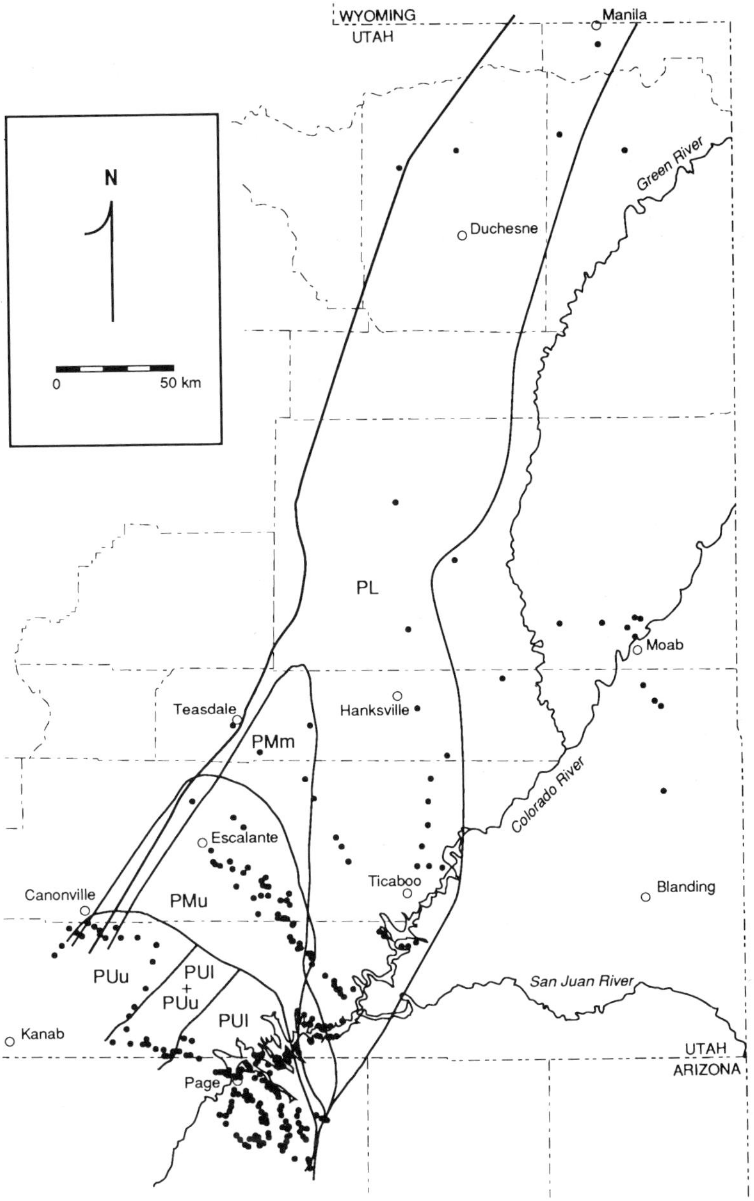

Fig. 6. Map showing the current geographic extent of major units in the Page Sandstone, with younger units covering a progressively smaller area. Units higher in the section (PMm, PMu, PUu) were originally more extensive and have been truncated by the uppermost (S--up) supersurface of the Page Sandstone.

GBS, Fig. 4D). This surface has irregular topography (Figs 3A & 4A, B, D & E). In some areas it is overlain directly by mudstones to siltstones of the Carmel Formation (e.g. east of Judd Hollow at section CME, Fig. 4D), but in most areas it is overlain by a unit that is transitional between the main aeolian body of the Page Sandstone and the mostly non-aeolian Carmel Formation.

Transition Unit (PTp, PTb)

In much of the southern part of the outcrop area of the Page Sandstone, there is an uppermost unit (PTp, PTb) that is typically 1 m thick or less, but in some areas is as much as 10 m thick (Fig. 4A, B & D). This Transition Unit consists of two lithofacies. The most geographically extensive facies (PTp) is a sandstone to silty sandstone facies consisting variably of wavy laminations, adhesion-ripple strata, subaqueous-ripple strata, contorted strata and thin silty to muddy drapes deformed locally into flame structures. In some areas this unit has on its upper surface raised ridges up to 1 m in height arranged in polygonal patterns with polygon diameters up to 15 m that appear to be large teepee structures (Havholm, 1991). The other lithofacies, which has been recognized in the Lake Powell area east of Antelope Island, consists of build-ups at the top of the Page Sandstone and associated detached sand bodies that overlie the transition zone (PTb, Figs 3A & 4B, D & E). The build-ups consist of linear to curvilinear ridges of sandstone ranging up to 12 m high, 0.5 km wide and 5 km long, are oriented generally southwest–northeast and occur both in isolation and in groups with regular spacing of the order of 0.1–0.2 km. The cores of the build-ups consist mainly of high-angle aeolian strata and are typically separated from the main body of the Page Sandstone by up to several metres of non-aeolian strata of the other facies of the Transition Unit (PTp). Strata in the build-ups commonly show evidence of slumping and fluidization and may have flanking drapes of mixed aeolian and subaqueous ripple strata. There are several examples of partly preserved dune forms with subaqueous ripple forms on their lower lee slopes. The detached sand bodies, which are poorly exposed, occur adjacent and parallel to the build-ups and in some exposures lap onto the build-ups. They appear to have dimensions of several tens of metres by several hundred metres in plan form and thicknesses up to 12 m, and to contain both wind-ripple and subaqueous-ripple strata.

The Transition Unit represents two events: a period of reworking of the uppermost part of the Page Sandstone after the erosional period represented by the UP supersurface, and one of aeolian reworking of coeval Carmel fluvial deposits (Blakey & Parnell, in press). The bulk of this unit (PTp) resulted from accumulation in sabkha to shallow subaqueous environments, but aeolian dunes did build up in some areas (PTb). These resulting dune strata were later partly reworked as they were submerged, forming in some areas the stacked sandstone elements of unit Jtb. Additionally, sandstone stringers with some build-ups (CS) lie above the Page encased within Carmel strata (Fig. 4B & D).

SUMMARY AND CONCLUSIONS

Careful tracing of supersurfaces, combined with recognition of the various aeolian and related stratification types, has allowed the distinction of six major sandstone packages within the Page Sandstone. These comprise a basal sabkha/sand sheet to fluviolacustrine unit (PB), a lower erg unit (PL), a middle amalgamated erg unit (PM), an upper erg unit (PU), an uppermost unit transitional from the aeolian Page Sandstone to the non-aeolian Carmel Formation (PT) and a unit found in outliers east of the main body of the Page Sandstone (PI). Some of these major packages are themselves made up of two or more supersurface bounded aeolian units. This illustrates that supersurfaces, which represent hiatal periods in erg accumulation, can be used to define a genetic stratigraphic framework within an aeolian sandstone on a regional basis. Because supersurfaces represent pauses in aeolian accumulation, which in turn represent changes in the aeolian sediment budget in the area, each supersurface bounded sandstone package represents a genetic unit, a distinct aeolian accumulation event. Moreover, bentonite layers, pods and drapes on dune foresets have been found in all units of the Page Sandstone except the Basal Unit. Such bentonites are also ubiquitous in the Carmel Formation, within which they have been radiometrically dated, and correlated on the basis of zircon morphologies (Everett *et al.*, 1990). This suggests that an absolute time frame may be added to the genetic stratigraphic framework of the Page Sandstone.

General theoretical discussions of the controls on sequence architecture have not included aeolian sandstones (e.g. Posamentier *et al.*, 1988; Galloway, 1989), partly because we have not understood the basin-scale process and response of aeolian systems very well. However, some progress has been made in understanding controls on aeolian stratigraphic architecture and how they differ from those of marine and fluvial units (e.g. Kocurek & Havholm, 1993). Several workers have begun to note the position of aeolian units relative to basinal sequence architecture (e.g. Kamola, 1989; Marzolf, 1990; Atchley & Loope, this volume). This study shows that aeolian events can now be recognized with great detail in the stratigraphic record. Integration of these several approaches should allow us to move towards integration of aeolian systems with systems of other sedimentary environments in basin sequence stratigraphic analysis.

ACKNOWLEDGEMENTS

The staff of the National Park Service at Glen Canyon National Recreation Area and at Arches National Monument helped in many ways to facilitate this study. The Navajo Nation provided several permits allowing access to Navajo land and this co-operation is greatly appreciated. Funding was provided by NSF grants EAR-8719972 and the Organized Research Committee, Northern Arizona University. N. Lancaster and J. Marzolf kindly reviewed the manuscript.

APPENDIX

Table A1. List of sections (section locations in italics are in Arizona, the others are in Utah)

	Section name	By	Location
AET	Antelope Tank	DK	*SE1/4, S16, T39N, R10E*
AIE	Antelope Island Northeast	KH	*SW1/4, S34, T42N, R9E*
AIN	Antelope Island North	RB	*SW1/4, S34, T42N, R9E*
AIW	Antelope Island West	RB	*SW1/4, S7, T41N, R9E*
ANC	Antelope Canyon (1) (ACN)	DK	*SW1/4, S23, T41N, R9E*
	Antelope Canyon (2)	DK	*NE1/4, S26, T41N, R9E*
APT	Airport (1)	MC	SE1/4, S21, T35S, R3E
	Airport (2)	MC	SE1/4, S22, T35S, R3E
	Airport (3)	MC	SE1/4, S16, T35S, R3E
ARC	Answer Canyon	DK	*SE1/4, S27, T40N, R12E*
ARM	Arrowhead Mesa	DK	*NW1/4, S6, T36N, R12E*
ARN	Arches North	KH	SW1/4, S3, T25S, R21R
ARS	Arches South	KH	SW1/4, S15, T25S, R21E
ATC	Antelope Creek (1)	DK	*SW1/4, S8, T39N, R10E*
	Antelope Creek (2)	DK	*NE1/4, S17, T39N, R10E*
	Antelope Creek (3)	DK	*NW1/4, S5, T39N, R10E*
	Antelope Creek (4)	DK	*NE1/4, S26, T40N, R9E*
	Antelope Creek (5)	DK	*SE1/4, S9, T39N, R10E*
	Antelope Creek (6)	DK	*NW1/4, S18, T39N, R10E*
	Antelope Creek (7)	DK	*SE1/4, S5, T39N, R10E*
	Antelope Creek (8)	DK	*NE1/4, S31, R10E, T40N*
	Antelope Creek (9)	DK	*NE1/4, S30, T40N, R10E*
AVC	Averett Creek	LJ	E1/2, S22, T38S, R2W
BCH	Bridger Canyon Head	KH, LJ	S1/2, S32, T43S, R1E
BDC	Boulder Cemetery	MC, KH	SE1/4, S20, T33S, R4E
BGN	Bull Valley Gorge North	LJ	N1/2, S28, T38S, R3W
BGP	Big P	KH, DK	*SW1/4, S12, T40N, R8E*
BGS	Bull Valley Gorge South	LJ	NW1/4, S27, T38S, R3W
BHA	Brimhall Arch	KH	SW1/4, S32, T35S, R9E
BHW	Big Hollow Wash (1a)	KH	SE1/4, S6, T39S, R7E
	Big Hollow Wash (2)	KH	NW1/4, S5, T39S, R7E
	Big Hollow Wash (3)	KH	SE1/4, S32, T38S, R7E
BKW	Buckhorn Wash	KH	NW1/4, S19, T19S, R11E
BMS	Bitman Spring (1)	DK	*SW1/4, S28, T40N, R12E*
	Bitman Spring (2)	DK	*SW1/4, S12, T39N, R11E*
BPO	Burr Point Overlook	KH	SE1/4, S26, T30S, R13E
BPT	Bridger Point	KH, LJ	SW1/4, S4, T44S, R1E
BRC	Balanced Rock Canyon	KH	SW1/4, S28, T42S, R8E
BRT	Burr Trail	MC, KH	W1/2, S15, T34S, R8E
BTD	Buck Tank Draw (L, U)	KH	SE1/4, S8 + NW1/4, S17 + SE1/4, S18, T43S, R2E
BTR	Behind the Rocks	KH	SE1/4, S21, T27S, R22E
BTS	Bentonite Section	KH	*SW1/4, S13, T41N, R8E*
BVD	Butler Valley Draw	LJ	1/4 miles N of Jcn Butler Valley Draw and Cottonwood Creek, T39S, R1W
BYF	Buoy Fifteen	KH	*SW1/2, S4, T44S, R5E*
CHR	Courthouse Rock	KH	NW1/4, S28, T24S, R20E
CKB	Creek Bed	DK	NW1/4, S6, T38N, R10E

Table A1. (continued)

	Section name	By	Location
CKS	Cockscomb	LJ	SE1/4, S11, T43S, R2W
CLC	Chaol Canyon (1)	DK	*SE1/4, S31, T39N, R12E*
	Chaol Canyon (2)	DK	*SE1/4, S5, T38N, R12E*
	Chaol Canyon (3)	DK	*NW1/4, S22, T38N, R12E*
CME	Cedar Mountain East	KH, LJ	SE1/4, S4, T44S, R2E
CPS	Cat Pasture	MC	SE1/4, S23, T38S, R6E
CPT	Cave Point (1)	KH	NW1/4, S35, T40S, R8E
	Cave Point (2)	KH	NW1/4, S35, T40S, R8E
CRW	Capitol Reef Waterfall	MC, KH	NE1/4, S15, T29S, R7E
CSC	Catstairs Canyon	LJ	SE1/4, S25, R2W, T42S
CTC	Cottonwood Canyon	LJ	W1/2, S3 + E1/2, S4, T41S, R1W
CWC	Circular White Cliffs (1)	DK	*SW1/4, S25, T38N, R9E*
	Circular White Cliffs (2)	DK	*NE1/4, S20, T38N, R9E*
	Circular White Cliffs (3)	DK	*NE1/4, S30, T38N, R10E*
	Circular White Cliffs (4)	DK	*NE1/4, S20, T38N, R10E*
	Circular White Cliffs (5)	DK	*NW1/4, S17, T38N, R10E*
DBA	Double Arch (E, C, W, WE)	KH	NW1/4, S30, T24S, R22E
DFC	Dry Fork Coyote (1)	KH	E1/2, S26, T38S, R6E
	Dry Fork Coyote (1a)	KH	W1/2, S25, T38S, R6E
	Dry Fork Coyote (2)	KH	S1/2, S25, T38S, R6E
DGC	Dungeon Canyon	RB, KH	NW1/4, S8, T43S, R7E
DMR	Dead Monkey Ridge	KH, DK	*SW1/4, S9, T38N, R9E*
DRC	Dangling Rope Canyon	KH	SW1/4, S30, T42S, R8E
DRN	Dry Rock Creek North	RB	S1/2, S14, T42S, R7E
DRS	Dry Rock Creek South	KH	SE1/4, S22, T42S, R7E
DWC	Driftwood Canyon	KH	S1/2, S27, T42S, R8E
ECH	Echo Cliffs (1)	DK	*SE1/4, S31, T40N, R8E*
	Echo Cliffs (2)	DK	NE1/4, S12, T39N, R8E
	Echo Cliffs (3)	DK	NE1/4, S14, T39N, R8E
	Echo Cliffs (4)	DK	NE1/4, S16, T39N, R8E
	Echo Cliffs (5)	DK	NW1/4, S26, T39N, R8E
EGP	Egypt (1)	MC	SE1/4, S35, T36S, R6E
	Egypt (2)	MC	SE1/4, S10, T37S, R6E
	Egypt (3)	MC	NE1/4, S15 + NW1/4, S14, T37S, R6E
ERB	Eighty-nine Road Bend	KH	*NE1/4, S16, T39N, R8E*
ESD	Escalante Dump	MC	NW1/4, S16, T35S, R3E
ETP	Edge of Thousand Pockets	RB, KH	*SW1/4, S24, T41N, R7E*
EWB	Early Weed Bench (1)	KH	NE1/4, S1, T38S, R6E
	Early Weed Bench (2)	KH	NW1/4, S1, T38S, R6E
	Early Weed Bench (3)	KH	SE1/4, S35, T37S, R6E
	Early Weed Bench (5)	KH	SE1/4, S34, T37S, R6E
	Early Weed Bench (6)	KH	SE1/4, S33, T37S, R6E
	Early Weed Bench (M)	MC	NW1/4, S1, T38S, R6E
FBN	Face Bay North (1, 2)	RB	N1/2, S35, T43S, R6E
FBS	Face Bay South	RB	S1/2, S36, T43S, R6E
FCC	Face Canyon (A)	KH	*NW1/4, S31, T42N, R11E*
	Face Canyon (B)	KH	*NW1/4, S31, T42N, R11E*
	Face Canyon (C)	KH	*SE1/4, S5, T41N, R11E*
	Face Canyon (D)	KH	*SW1/4, S33, T42N, R11E*
	Face Canyon (E)	KH	S1/2, S3, T44S, R6E
	Face Canyon (F)	KH	*E1/2, S32, T42N, R11E*
	Face Canyon (G)	KH	NE1/4, S11, T44S, R6E
	Face Canyon (H)	KH	NW1/4, S11, T44S, R6E
	Face Canyon (I)	KH	SE1/4, S3, T44S, R6E
	Face Canyon (J)	KH	SE1/4, S3, T44S, R6E
FCL	Face Cliffs (N, W, SW, E)	RB	NW1/4, S3, T44S, R6E
FCW	Face Canyon West (1–3)	RB	E1/2, S5, T44S, R6E
FMR	Fiftymile Ridge	KH	SE1/4, S11, T41S, R81/2E
FMW	Fiftymile Wash	KH	NE1/4, S12, T41S, R8E

Table A1. (continued)

	Section name	By	Location
FSV	Ferry Swale View	KH	*NE1/4, S20, T41N, R8E*
FTH	Fourth Valley	KH	SW1/2, S2, T44S, R1E
FTC	Fault Mesa	DK	*SW1/4, S24, T39N, R11E*
FWD	Four Wheel Drive	DK	*SE1/4, S6, T38N, R10E*
FWM	Flat Wide Mesa	DK	*SW1/4, S24, T40N, R8E*
GBN	E. Gregory Bay North	RB	NW1/4, S13, T43S, R61/2E
GBS	E. Gregory Bay South	RB	NW1/4, S24, T43S, R61/2E
GBV	Goblin Valley	KH	SE1/4, S11, T25S, R11E
GCN	Grotto Canyon North	RB	NE1/4, S5, T43S, R7E
GCS	Grotto Canyon South	RB	SE1/4, S5, T43S, R7E
GDE	Garden of Eden	KH	SW1/4, S24, T24S, R21E
GHO	Good Hope Mesa Overlook	KH	SE1/4, S36, T351/2S, R12E
GLP	Gravel Pit	DK	*NE1/4, S13, T40N, R8E*
GWE	Grotto-Wetherill East	KH	SW1/4, S33, T421/2S, R7E
GWW	Grotto-Wetherill West	KH	SW1/4, S33, T421/2S, R7E
HAN	Hanna	KH	Ctr, S1/2, S31, T1N, R7W
HBC	Hackberry Canyon	LJ	E1/2, S31, T38S, R1W
HBT	Hackberry Tank	KH	*NE1/4, S21, T41N, R8E*
HCB	Halls Crossing Beach	KH	Ctr, S32, T38S, R11E
HCM	Halls Crossing Maintenance	KH	NE1/4, S32, T381/2S, R11E
HCN	Halls Creek North	RB	SW1/4, S23, T38S, R10E
HCS	Halls Creek South	KH	SE1/4, S25, T38S, R10E
HHS	Hell Hole Swale	KH	NW1/4, S27, T28S, R12E
HIR	Hole-in-the-Rock	KH	NE1/4, S33, T401/2S, R9E
HSC	Horseshoe Canyon	KH	SE1/4, S7, T27S, R16E
HTM	Horsethief Mesa	KH, DK	*NW1/4, S2, T37N, R11E*
HWA	Hurricane Wash (A)	KH	SW1/4, S26, T39S, R7E
HWH	Halfway Hollow (1, 2)	MC	NW1/4, S35, T36S, R4E
HWL	Hurricane Wash (L)	KH	SW1/4, S26, T39S, R7E
HWS	Harris Wash (1)	MC	SE1/4, S33, T36S, R5E
	Harris Wash (2)	MC	NE1/4, S5, T36S, R5E
HYD	Honey Draw	DK	*NE1/4, S18, T40N, R9E*
HZD	Hazdi	DK	*NE1/4, S9, T40N, R9E*
JCT	Jacob's Tanks (N, S)	KH	SE1/4, S15, T43S, R2E
JHB	Judd Hollow Bend	KH, LJ	SW1/4, S31, T43S, R2E
JHE	Judd Hollow East	KH, LJ	E1/2, S36, T43S, R1E
JHL	Judd Hollow Power Lines	KH, LJ	SE1/4, S35, T43S, R1E
JKN	Julia's Knob	JK	*NE1/4, S24, T41N, R8E*
JNC	Junction	KH	NW1/4, S14, T32S, R12E
KBT	Kaibito	KH	*NE1/4, S29, T37N, R12E*
KSP	Kane Springs	KH	SW1/4, S12, T28S, R22E
LBE	Labyrinth Canyon East	RB	NE1/4, S7, T44S, R6E
LBW	Labyrinth Canyon West	RB	S1/2, S1, T44S, R5E
LCA	Labyrinth Canyon (A)	KH	NE1/4, S2, T41N, R10E
LCB	Labyrinth Canyon (B)	KH	*NE1/4, S2, T41N, R10E*
LCE	Lake Canyon East	KH	NW1/4, S30, T38S, R12E
LCM	Labyrinth Canyon Mouth	RB, KH	E1/2, S6, T44S, R6E
LCR	Leche-e Rock (1) (LCK)	KH, DK	*NW1/4, S13, T40N, R10E*
	Leche-e Rock (2)	DK	*NW1/4, S3, T40N, R10E*
	Leche-e Rock (3)	DK	*SW1/4, S33, T41N, R10E*
	Leche-e Rock (4)	DK	*SE1/4, S33, T41N, R10E*
	Leche-e Rock (5)	DK	*NW1/4, S19, T41N, R10E*
	Leche-e Rock (6)	DK	*SW1/4, S18, T41N, R10E*
	Leche-e Rock (7)	DK	*NE1/4, S18, T41N, R10E*
	Leche-e Rock (8)	DK	*SE1/4, S29, T41N, R10E*
	Leche-e Rock (9)	DK	*SE1/4, S2, T40N, R10E*
	Leche-e Rock (10)	DK	*NE1/4, S11, T40N, R10E*
	Leche-e Rock (11)	DK	*SE1/4, S30, T40N, R11E*
	Leche-e Rock (12)	DK	*SW1/4, S31, T40N, R11E*

Table A1. (continued)

	Section name	By	Location
LCW	Lake Canyon West	KH	NE1/4, S34, T381/2S, R11E
LDR	Long Draw	LJ	SW1/4, S8, T43S, R1W
LFK	Lake Fork	KH	NE1/4, S3, T1N, R5W
LKW	Lick Wash	LJ	NE1/4, S1, T40S, R4W
LNM	Lion Mountain	MC, KH	NW1/4, S26 + NE1/4, S27, T30S, R5E
LPC	Lower Podunk Creek	LJ	SW1/4, S30, T39S, R3W
LPO	Lake Powell Overlook	KH	NE1/4, S3, T44S, R2E
LSF	Little Spencer Flat (1)	MC	N1/2, S29, T35S, R4E
	Little Spencer Flat (2)	MC	NW1/4, S35, T35S, R4E
MAN	Manila	KH	SW1/4, S6, T2N, R20E
MEF	Mile Eighty-five	MC	SE1/4, S35, T33S, R4E
MDS	Middle Section	KH	*NW1/4, S22, T41N, R8E*
MHN	Mount Holmes North	KH	NW1/4, S28, T34S, R12E
MKF	Marker Fifteen	KH	S1/2, S4, T44S, R5E
MMN	Manson Mesa N	DK	*SE1/4, S19, T41N, R9E*
MMS	Manson Mesa South	DK	*NW1/4, S31, T41N, R9E*
MMW	Manson Mesa NW (MNW)	KH, DK	*NW1/4, S19, T41N, R9E*
MNE	Manson Mesa NE	KH, DK	*E1/2, S20, T41N, R9E*
MNR	Mormon Ridge (1)	DK	*NE1/4, S8, T36N, R11E*
	Mormon Ridge (2)	DK	*SW1/4, S11, T36N, R11E*
MNW	Manson Mesa NW (MMW)	KH, DK	*NW1/4, S19, T41N, R9E*
MRN	Middle Rock Creek North	RB	NE1/4, S16, T42S, R7E
MRS	Middle Rock Creek South	KH	NE1/4, S21, T42S, R7E
MSC	Muleshoe Canyon	KH	S1/2, S18, T28S, R23E
MTS	Mountain Sheep	RB	NW1/4, S8, T43S, R8E
MWC	Maidenwater Creek	KH	SW1/4, S26 + NW1/4, S35, T33S, R12E
NCN	Navajo Canyon	KH	*NE1/4, S16, T41N, R10E*
NHB	New Home Bench	MC	SW1/4, S35, T33S, R4E
NMC	Nightmare Corner	KH	*SW1/4, S23, T41N, R8E*
OCC	Oak Creek Canyon	MC, KH	NE1/4, S30 + NW1/4, S29, T31S, R7E
PAP	Page Airport	DK	*SE1/4, S20, T41N, R9E*
PCK	Pine Creek	MC	SW1/4, S29, T34S, R3E
PHG	Photograph Gap	KH	NE1/4, S29, T31S, R23E
PLR	Power Line Road	DK	*NE1/4, S19, T39N, R11E*
PRN	Paria River North	LJ	NW1/4, S29, T38S, R2W
PRV	Paria River View	KH	NE1/4, S23, T43S, R1W
PTK	Poulas Tanks	KH	SE1/4, S2, T36S, R11E
PTM	Peanut Mesa (1)	DK	*NE1/4, S33, T39N, R11E*
	Peanut Mesa (2)	DK	*SE1/4, S21, T39N, R11E*
	Peanut Mesa (3)	DK	*NE1/4, S21, T39N, R11E*
	Peanut Mesa (4)	DK	*NW1/4, S34, T39N, R11E*
PWA	Paria Wilderness Area	KH, LJ	NW1/4, S14, T43S, R1W
QNC	Question Canyon	DK	*NE1/4, S35, T40N, R12E*
RBK	Red Breaks (1)	MC	SW1/4, S14 + NW1/4, S23, T36S, R5E
	Red Breaks (2)	MC	E1/2, S15, T36S, R5E
	Red Breaks (3)	MC	N1/2, S16, T36S, R5E
RBM	Road Bend Mesa	KH, DK	*S Ctr S19, N Ctr S30, T38N, R11E*
RBO	Road Bend 160	DK	*SW1/4, S8, T38N, R9E*
RCW	Rock Creek West	RB	Middle, S6, T42S, R7E
RDC	Red Canyon	KH, MC	S1/2, S23, T32S, R7E
RDR	Red Ridge (1)	DK	*NE1/4, S29, T37N, R12E*
	Red Ridge (2)	DK	*SE1/4, S32, T38N, R12E*
	Red Ridge (3)	DK	*NW1/4, S8, T37N, R12E*
RDT	Radio Towers	KH	*SE1/4, S23, T41N, R8E*
RDV	Red Valley	KH	NW1/4, S4, T44S, R2E
RFC	Radio Facility	MC	SW1/4, S24, T35S, R4E
RHC	Rawhide Cave	KH	*SE1/4, S20, T41N, R8E*
RPN	Rainbow Plateau North	KH	*S1/2, S35, T42N, R9E*
RPS	Rainbow Plateau South	RB	*S1/2, S35, T42N, R9E*

Table A1. (continued)

	Section name	By	Location
RPT	Red Point	KH	*SE1/4, S22, T41N, R8E*
RRT	Railroad Track (1)	DK	*NE1/4, S34, T39N, R10E*
	Railroad Track (2)	DK	*SE1/4, S3, T38N, R10E*
	Railroad Track (3)	DK	*SE1/4, S23, T38N, R10E*
RSB	Rush Beds	LJ	NW1/4, S1, T40S, R1W
RSP	Rock Springs	LJ	SE1/4, S22, T38S, R2W
RVD	Round Valley Draw	LJ	SW1/4, S33, T38S, R1W
RWL	Red Well (1)	KH	SW1/2, S10 + NW1/2, S15, T39S, R7E
	Red Well (2)	KH	SE1/4, S10, T39S, R7E
	Red Well (2d)	KH	SW1/4, S11, T39S, R7E
	Red Well (3)	KH	NW1/4, S14, T39S, R7E
SDP	Sewage Disposal Pond	DK	*NW1/4, S15, T40N, R9E*
SDV	Sand Valley	KH	*SW1/4, S34, T42N, R7E*
SEC	Second Valley	KH	NE1/4, S2, T44S, R1E
SFL	Seep Flat	MC	Middle, S6, T37S, R5E
SFT	South Fork Ticaboo	KH	SW1/4, S27, T351/2S, R12E
SHD	Saharan Dunes	KH	NW1/4, S24, T31S, R12E
SHL	Sand Hills (2, 4)	RB	SE1/4, S34, T43S, R5E
	Sand Hills (3)	RB	SW1/4, S35, T43S, R5E
SRD	Shooting Range	DK	*NE1/4, S5, T40N, R9E*
SRK	Shooting Range	KH	*S1/2, S4, T40N, R9E*
SLM	Salt Lake Meridian	KH, LJ	SW1/4, S30, T43S, R1E
SNG	Sooner Gulch (1)	KH	N1/2, S26, T40S, R8E
	Sooner Gulch (2)	KH	NE1/4, S26, T40S, R8E
	Sooner Gulch (4)	KH	NW1/4, S26, T40S, R8E
SNW	Sooner Wash (1, 1a)	KH	NW1/4, S16, T40S, R8E
	Sooner Wash (2)	KH	NE1/4, S16, T40S, R8E
SPC	Sheep Creek	LJ	SE1/4, S24, T38S, R3W
SPM	Steep Mesa	DK	SE1/4, S15, T40N, R9E
SPP	Steep Point	DK	SE1/4, S28, T40N, R9E
SSP	Soda Spring (1)	KH	NE1/4, S2, T41S, R8E
	Soda Spring (2)	KH	NW1/4, S1, T41S, R8E
STR	Steinaker	GK	S36, T7N, R21E
SWW	Starting Water Wash	DK	SE1/4, S30, T39N, R12E
TBX	The Box	MC	NW1/4, S12, T34S, R2E
TDB	Tidwell Bottoms	KH	SW1/4, S6, T22S, R14E
TFM	Twentyfive Mile Wash (1)	MC	SW1/4, S29, T37S, R6E
	Twentyfive Mile Wash (2)	MC	SW1/4, S32, T37S, R6E
TMF	Tenmile Flat (1–3)	MC	NW1/4, S16, T36S, R4E
TND	The Needles	KH	W1/2, S27, T24S, R18E
TPS	The Post	MC, KH	W1/2, S25, T34S, R8E
TSD	Teasdale	MC, KH	SE1/4, S17, T29S, R4E
TSG	Tse Esgizzi (1)	DK	*NE1/4, S20, T39N, R12E*
	Tse Esgizzi (2)	DK	*SE1/4, S13, T39N, R11E*
	Tse Esgizzi (3)	DK	*SW1/4, S12, T39N, R11E*
TSL	Tse Lichii (2)	DK	*SW1/4, S30, T40N, R12E*
TWC	Twilight Canyon	KH	NE1/4, S23, T42S, R81/2E
TMW	Twentymile Wash	MC	NW1/4, S28, T37S, R5E
UNM	Unnamed Mesa (1)	MC	SW1/4, S3, T36S, R5E
	Unnamed Mesa (2)	MC	NW1/4, S9, T36S, R5E
VWP	View Point (1)	DK	*SW1/4, S5, T39N, R11E*
	View Point (2)	DK	*SW1/4, S8, T39N, R11E*
WCB	West Clark Bench	KH	NE1/4, S22, T43S, R1W
WCE	West Clark Bench East	LJ	SW1/4, S28 + NE1/4, S29, T43S, R1W
WCN	Willis Creek North	LJ	NW1/4, S23, T38S, R3W
WCS	Willis Creek South	LJ	SE1/4, S23, T38S, R3W
WCW	West Clark Bench West	LJ	SW1/4, S36, T43S, R1W
WHC	Water Holes Canyon	KH	*SW1/4, S25, T40N, R8E*
WHT	Wild Horse Trap Mesa	KH	*SE1/4, S16, T38N, R10E*

Table A1. (continued)

	Section name	By	Location
WHV	Water Hole Valley (1)	DK	*NE1/4, S9, T38N, R11E*
	Water Hole Valley (2)	DK	*NW1/4, S16, T38N, R11E*
	Water Hole Valley (3)	DK	*SW1/4, S8, T38N, R11E*
WLT	Willow Tank	KH	*SE1/4, S19, T41N, R8E*
WPP	Wooden Power Poles	KH	*NE1/4, S21, T41N, R8E*
WRK	Whiterocks	KH	SW1/4, S8 + NW1/4, S17, T2N, R1E
WLC	Wetherill Canyon	RB	SW1/4, S13, T43S, R7E
WST	West Canyon (1)	KH	NE1/4, S23, T43S, R61/2E
	West Canyon (2)	KH	SE1/4, S24, T43S, R61/2E
WSW	West Canyon W	RB	NE1/4, S36, T43S, R6E

DK, Dave King; GK, Gary Kocurek; JK, Julia Knight; KH, Karen Havholm; LJ, Larry Jones; MC, Mark Capps; RB, Ron Blakey.

REFERENCES

ANTON, D. (1983) Modern eolian deposits of the eastern province of Saudi Arabia. In: *Eolian Sediments and Processes* (Ed. Brookfield, M.E. & Ahlbrandt, T.S.) pp. 379–406. Developments in Sedimentology 38. Elsevier, Amsterdam.

BLAKEY, R.C. (1988) Superscoops: their significance as elements of eolian architecture. *Geology* **16**, 483–487.

BLAKEY, R.C. & PARNELL, R.A., JR (in press) Middle Jurassic magmatism: the volcanic record in the eolian Page Sandstone and sabkha–marine Carmel Formation, Colorado Plateau. In: *Jurassic Magmatism and Tectonics of the North American Cordillera* (Ed. Miller, D. & Busby, C.J.). Geol. Soc. Am. Memoir.

BLAKEY, R.C., PETERSON, F., CAPUTO, M.V., GEESAMAN, R.C. & VOORHEES, B.J. (1983) Paleogeography of Middle Jurassic continental, shoreline, and shallow marine sedimentation, southern Utah. In: *Mesozoic Paleogeography of the West-central US* (Ed. Reynolds, M.W. & Dolly, E.D.) pp. 77–100. Rocky Mountain Paleogeography Symposium 2. Rocky Mtn Section, Soc. Econ. Paleontol. Mineral., Denver.

BLAKEY, R.C., PETERSON, F. & KOCUREK, G. (1988) Synthesis of late Paleozoic and Mesozoic eolian deposits of the Western Interior of the United States. In: *Late Paleozoic and Mesozoic Eolian Deposits of the Western Interior of the United States* (Ed. Kocurek, G.) pp. 3–125. Sed. Geol. 56.

CAPPS, M. (1990) *Presence and significance of regional bounding surfaces and genetic sequences in an eolian sandstone: Page Sandstone (Jurassic), south-central Utah*. Unpublished MS thesis, Northern Arizona University, Flagstaff.

CAPUTO, M.V. (1980) *Depositional history of Middle Jurassic clastic shoreline sequences in southwestern Utah*. Unpublished MS thesis, Northern Arizona University.

DEYNOUX, M., KOCUREK, G. & PROUST, J.-N. (1989) Late Proterozoic periglacial aeolian deposits on the West African Platform, Taoudeni Basin, western Mali. *Sedimentology* **36**, 531–549.

EVERETT, B.H., KOWALLIS, B.J., CHRISTIANSEN, E.H. & DEINO, A. (1990) Correlation of Jurassic sediments of the Carmel and Twin Creek Formations of southern Utah using bentonite characteristics. *Utah Geological and Mineral Survey Open File Report 169*.

FRYBERGER, S.G. & KOELMEL, M.H. (1986) Rangely Field: eolian system-boundary trap in the Permo-Pennsylvanian Weber Sandstone of northwest Colorado. *Rocky Mountain Association of Geologists Symposium*, 129–149.

GALLOWAY, W.E. (1989) Genetic stratigraphic sequences in basin analysis I: architecture and genesis of flooding-surface bounded depositional units. *Am. Ass. Petrol. Geol. Bull.* **73**, 125–142.

GEESAMAN, R.C. (1979) *Sedimentary facies of the Carmel Formation, southeast Utah*. Unpublished MS thesis, Northern Arizona University.

GLENNIE, K.W. (1970) *Desert Sedimentary Environments*. Elsevier, Amsterdam, 222 pp.

HAVHOLM, K.G. (1991) *Eolian event stratigraphy: theory, and application to the Middle Jurassic Page Sandstone, Colorado Plateau, USA*. Unpublished PhD thesis, University of Texas, Austin.

JONES, L.S. (1990) *Stratigraphy and depositional history of the Page Sandstone and correlative units of the Carmel Formation, Middle Jurassic, south-central Utah*. Unpublished MS thesis, Northern Arizona University, Flagstaff.

KAMOLA, D.L. (1989) Permian–Triassic boundary in south-central Utah: paleorelief and sequence stratigraphy [abstract]. *Geol. Soc. Am. Ann. Mtng, Abstracts with Programs* **21**, A80.

KING, D.D. (1992) *Stratigraphic analysis of the landward margin of the Middle Jurassic Page Sandstone, Coconino County, north-central Arizona*. Unpublished MS thesis, Northern Arizona University, Flagstaff.

KOCUREK, G. (1984) Origin of first-order bounding surfaces in aeolian sandstones—reply. *Sedimentology* **31**, 125–127.

KOCUREK, G. (1988) First order and super bounding surfaces in eolian sequences—bounding surfaces revisited. In: *Late Paleozoic and Mesozoic Eolian Deposits of the Western Interior of the United States* (Ed. Kocurek, G.) pp. 193–206. Sed. Geol. 56.

KOCUREK, G., DEYNOUX, M., BLAKEY, R.C. & HAVHOLM, K.G. (1991a) Amalgamated deposits resulting from climatic and eustatic changes, Akchar Erg, Mauritania. *Sedimentology* **38**, 751–772.

KOCUREK, G. & HAVHOLM, K.G. (1993) Eolian event stratigraphy—a conceptual framework. In: *Recent Advances in and Applications of Siliciclastic Sequence Stratigraphy* (Ed. Weimer, P. & Posamentier, H.). Am. Ass. Petrol. Geol. Memoir. Am. Ass. Petrol. Geol., Tulsa (in press).

KOCUREK, G. & HUNTER, R.E. (1986) Origin of polygonal fractures in sand, uppermost Navajo and Page Sandstones, Page, Arizona. *J. Sed. Petrol.* **56**, 895–904.

KOCUREK, G., KNIGHT, J. & HAVHOLM, K. (1991b) Outcrop and semi-regional three-dimensional architecture and reconstruction of a portion of the eolian Page Sandstone (Jurassic). In: *Three-dimensional Facies Architecture* (Ed. Miall, A. & Tyler, N.). Soc. Econ. Paleont. Mineral. Atlas pp. 25–43. Soc. Econ. Paleontol. Mineral., Tulsa.

LOOPE, D.B. (1985) Episodic deposition and preservation of eolian sands: a late Paleozoic example from southeastern Utah. *Geology* **13**, 73–76.

MARZOLF, J.E. (1990) Speculation on the sequence stratigraphic significance of eolian sandstones: examples from the western interior. *Abst. 13th International Sedimentological Congress, Nottingham*, 334.

MCCLURE, H.A. (1978) Ar Rub'al Khali. In: *Quaternary Period in Saudi Arabia* (Ed. Al-Sayari, S.S. & Faure, J.G.) pp. 252–263. Springer-Verlag, Wien.

PETERSON, F. (1986) Jurassic paleotectonics in the west-central part of the Colorado Plateau, Utah and Arizona. In: *Paleotectonics and Sedimentation, Rocky Mountain Region, US* (Ed. Peterson, J.A.) pp. 563–596. Am. Ass. Petrol. Geol. Spec. Mem. 41. Am. Ass. Petrol. Geol., Tulsa.

PETERSON, F. & PIPIRINGOS, G.N. (1979) Stratigraphic relations of the Navajo Sandstone to Middle Jurassic formations, southern Utah and northern Arizona. *US Geol. Surv. Prof. Pap. 1035B*, B1–B43.

POSAMENTIER, H.W., JERVEY, M.T. & VAIL, P.R. (1988) Eustatic controls on clastic deposition I—conceptual framework. In: *Sea-level Changes: an Integrated Approach* (Ed. Wilgus, C.K., Hastings, B.S., Kendall, C.G.St C., Posamentier, H.W., Ross, C.A. & Van Wagoner, J.) pp. 109–124. Soc. Econ. Paleontol. Mineral. Spec. Pub. 42. Soc. Econ. Paleontol. Mineral., Tulsa.

SALEEBY, J. & BUSBY-SPERA, C. (1992) Early Mesozoic tectonic evolution of the western US Cordillera. In: *The Cordilleran Oregon: Conterminus US* (Ed. Burchfield, B.C. *et al.*). Geological Society of America, The Decade of North American Geology, G-3, pp. 107–168. Geol. Soc. Am., Boulder.

SWEZEY, C. (1991) *Description and interpretation of the Jurassic J-2 unconformity of the Western Interior (USA)*. Unpublished MA thesis, University of Texas, Austin.

TALBOT, M.R. (1985) Major bounding surfaces in aeolian sandstones—a climatic model. *Sedimentology* **32**, 257–265.

VAN WAGONER, J.C., POSAMENTIER, H.W., MITCHUM, R.M., VAIL, P.R., SARG, J.F., LOUTIT, T.S & HARDENBOL, J. (1988) An overview of the fundamentals of sequence stratigraphy and key definitions. In: *Sea-level Changes: an Integrated Approach* (Ed. Wilgus, C.K., Hastings, B.S., Kendall, C.G.St C., Posamentier, H.W., Ross, C.A. & Van Wagoner, J.) pp. 39–45. Soc. Econ. Paleontol. Mineral. Spec. Pub. 42. Soc. Econ. Paleontol. Mineral., Tulsa.

VOORHEES, B.J. (1978) *Stratigraphy and facies of the lower Carmel Formation (Middle Jurassic), southwestern Utah.* Unpublished MS thesis, Northern Arizona University, Flagstaff.

Spec. Publs Int. Ass. Sediment. (1993) **16**, 109–126

Downwind changes within an ancient dune sea, Permian Cedar Mesa Sandstone, southeast Utah

R.P. LANGFORD* *and* M.A. CHAN[†]

**Bureau of Economic Geology, University of Texas at Austin, Austin, TX 78712, USA; and [†]University of Utah, Salt Lake City, UT 84112, USA*

ABSTRACT

Stratigraphic sections along a wind-parallel transect within the upper part of the Permian Cedar Mesa Sandstone illustrate changes relative to the downwind erg margin. Six phenomena may help to define the erg margin: bioturbation, wide inter-dune areas, abundant sand sheets or zibars, abundance of non-aeolian facies, smaller and different types of aeolian dunes, and an increased abundance of erosion surfaces. Each phenomenon responds differently to stratigraphic and palaeogeomorphological changes. Bioturbation varies stratigraphically rather than laterally, and may respond to climatic or water table controls rather than to changes in position within the dune sea. Dune spacing and the sizes of each type of dune do not show much stratigraphic or lateral downwind change. The proportions of different dune types change regularly as the abundance of large-trough cross-stratified sets decreases downwind. The proportion of sand sheet deposits increases regularly downwind. Grain size decreases regularly downwind but does not indicate position relative to the erg margin. Non-aeolian facies increase towards the erg margins, but can also form a significant proportion well into the dune sea. Erosion surfaces increase in frequency at the downwind erg margin from one per 17 m of section in the five upwind sections to one per 9 m of section and offer the most dramatic indicator of marginality.

The consistent average grain size at each section indicates that the sources of aeolian sand did not significantly change either in position or in sand composition throughout the periods of deposition.

Individual deflation surface bounded stratigraphic units do not change in grain size upward, implying that (i) periods of erosion were not long-lived, or (ii) the sand supply was shut off during periods of deflation, either due to removal of the supply or through a change in the transporting wind direction.

INTRODUCTION

To what extent do changes in depositional features within aeolian sandstones indicate expansion, contraction or migration of the erg?

Some studies have emphasized the sensitivity of aeolian deposition to changes in climate and other influences (for discussion see Kocurek, 1988). Recent papers have documented differences between the interiors of modern sand dune seas and their margins. These papers suggest that erg margins may be more vegetated and more bioturbated; contain wider, more integrated inter-dunes; have abundant sand sheets or zibars; have extra-erg, non-aeolian environments such as sabkhas, playas and/or fluvial channels; and contain smaller and differently shaped aeolian bedforms. Temporal and spatial changes in features of ancient aeolianites have also been related to erg margins. The Permian Cedar Mesa Sandstone provides a unique opportunity to measure the spatial changes within sand dune seas because the margins of the dune sea were relatively stable through time. Thus differences within the Cedar Mesa may relate to spatial rather than temporal changes in aeolian sedimentation.

One objective of this paper is to summarize some of the current ideas on the recognition of spatial changes within modern and ancient ergs and erg margins. The second objective is to determine the applicability of these criteria to a well exposed ancient example from the Permian Cedar Mesa Sandstone. In particular, the paper describes the changes that occur in stratigraphic sections along a wind-parallel transect within the upper part of the Cedar Mesa Sandstone, and relates depositional features to position relative to the erg margins.

BACKGROUND

Spatial changes described from modern and ancient sand dune seas will be discussed in the following six categories:

1 grain size and sorting;
2 dune size and dune forms;
3 dune spacing;
4 abundance of sand sheets, zibars and inter-dunes;
5 abundance of bioturbation;
6 abundance of non-aeolian deposits.

For each of the categories numbered above, both modern and ancient examples are briefly cited to summarize the literature and to demonstrate the amount of diversity that has been observed in ergs and their margins.

Modern studies

Grain size and sorting

Grain size and sorting changes within ergs are described by numerous authors (Folk, 1971; Lancaster, 1982, 1986; Ash & Wasson, 1983; Wasson, 1983). Grain size has been described as a function of source sediment (Folk, 1971; Wilson, 1973; Lancaster, 1982, 1983; Wasson, 1983) and distance of transport (Wilson, 1973; Lancaster, 1986). Dunes become finer-grained and better sorted downwind, approaching a well sorted population of grains between 2 and 3 ϕ (Folk, 1971; Wilson, 1973; Lancaster, 1983, 1986; Wasson, 1983).

Two distinct types of ergs are differentiated according to whether sand sources are coarser, or alternatively finer, than fine-grained sand (Lancaster, 1986). Wilson (1973) and Lancaster (1982, 1983) described grain size trends in coarse-sourced ergs, while Wasson (1983), Lancaster (1986) and Folk (1971) described fine-sourced ergs. The primary differences between the two types are that coarse-sourced ergs have inter-dunes that are coarser-grained than the dunes, and also exhibit trends of decreasing grain size and increasing sorting upward on individual dunes and draas (Wilson, 1973; Lancaster, 1982, 1983). Grain size also decreases downwind (Wilson, 1973). Fine-sourced ergs have finer-grained inter-dunes, and also exhibit an increase in sorting with no decrease in grain size upward on individual dunes and draas (Lancaster, 1986).

Dune size and dune forms

Numerous authors have noted that dunes decrease in size and complexity at erg margins. Maps of the White Sands dunefield showed complex barchanoid dunes between domal dunes at the upwind margin and parabolic dunes at the downwind margin (McKee, 1966). Fryberger *et al.* (1979) and Andrews (1981) found small, simple coppice, parabolic and transverse dunes within the marginal sand sheet of the Great Sand Dunes of Colorado, compared to the complex transverse and star dunes within the dune mass. Lancaster (1983) mapped dune height in the Namib Sand Sea. Large and complex dunes were found at portions of the margins of the dune sea. However, dunes on the margins were generally smaller and simpler in form, particularly on the upwind margin. Barchanoid and transverse dunes were concentrated on the upwind margins while complex linear and star dunes formed the bulk of the dunefield including the downwind margin.

However, dune type may also change within the interior of dune seas. Lancaster (1983) illustrates the spatial changes from simple to complex linear dunes to star dunes within the Namib desert.

Dune spacing

Dune spacing varies in its relationships to erg margins. However, numerous other controls also affect dune spacing. Large, complex linear and star dunes within the same dune sea may respond to different processes than do smaller linear, transverse and barchanoid dunes (Lancaster, 1988). Commonly, linear and oblique dunes are more widely spaced at erg margins. Mainguet & Chemin (1983) showed inter-dune corridors widening at the margins of Saharan dunefields. Breed *et al.* (1979) showed several clearly photographed examples of increases in inter-dune areas on the margins of ergs.

However, in barchanoid, transverse or star dunefields, dune spacing decreases towards the margins. Lancaster (1983) mapped smaller dune spacing at most of the margins of the Namib Sand Sea. Dune spacing correlated with grain size and with dune height, and maximum dune spacings correlated with the largest dunes in the centre of the erg. Breed *et al.* (1979) furnished several examples where dune spacing and size decrease at erg margins. Wilson (1973) mapped dune height and spacing, and noted that these correlated with the percentage of area covered by sand and average sand thickness. He noted that thick erg deposits were associated with large, complex dunes and draas, whereas thin erg deposits were associated with smaller, simple dune forms. There was a drastic decrease in dune size and spacing at the erg margins.

Inter-dunes are more variable than their accompanying aeolian deposits and record environmental conditions of deposition (Kocurek, 1981). Some authors have described increases in the inter-dune area at the margins of dunefields. Although individual inter-dunes are smaller, they cover a larger proportional area. McKee (1966) noted that inter-dune areas covered a larger percentage of total area near the margins of the White Sands dunefield. Wilson (1973) mapped sand coverage in Saharan ergs. The percentage of sand coverage was lowest at the erg margins. Dune spacing correlated with the size of the coarsest 5% of sand grains on the dunes.

Abundance of sand sheets, zibars and inter-dunes

Wide, low-relief depositional areas also characterize erg margins. Three low-relief aeolian features that have been described are zibars, sand sheets and inter-dunes. Nielson & Kocurek (1986) documented zibars, coarse-grained, low-relief aeolian bedforms marginal to the Algodones dunefield in southern California. Kocurek & Nielson (1986) documented extensive sand sheets on the margins of the Great Sand Dunes, Algodones and Gran Desierto dunefields. Periodic flooding, vegetation and a coarse-grained sand supply were inferred to favour sand sheet development.

Breed *et al.* (1979) illustrated the global distribution and common occurrence of sand sheets on the margins of ergs. Fryberger *et al.* (1979) described extensive sand sheet deposits adjacent to the Great Sand Dunes. The low-angle, cross-stratified sand sheet sands were interpreted as important erg margin deposits. Gently dipping fine- and coarse-grained laminae were common within these deposits. Fryberger *et al.* (1983) observed sand sheets transitional between aeolian and non-aeolian facies in a dune sea in eastern Saudi Arabia. Sand sheets were interpreted as a residual facies, formed after the passage of many dunes when a protective mantle of vegetation and coarse grains formed. Flat, siliciclastic sabkhas were concentrated on the seaward margin of the Saudi dune sea (Fryberger *et al.*, 1983).

Abundance of bioturbation

Although vegetation in dunefields within a given region is a function of climate (Ash & Wasson, 1983), increased vegetation is common to the margins of many dune seas in semi-arid climates. Fryberger *et al.* (1979) and Andrews (1981) described vegetation in a sand sheet adjacent to the Great Sand Dunes. Kocurek & Nielson (1986) documented vegetated sand sheets on the margins of several dunefields in western North America. McKee (1966) described and photographed vegetated parabolic dunes and inter-dunes on the downwind margin of the White Sand Dunes. Bioturbation from both plants and animals is more intense at the margins of dunefields where the rate of dune migration may be slowed by vegetation (Fryberger *et al.*, 1979).

Abundance of non-aeolian deposits

Glennie (1970) described inter-bedded fluvial and aeolian deposits from the margins of dune seas in Saudi Arabia and noted the importance of non-aeolian deposits. Ahlbrandt & Fryberger (1981) created a model in which dunefields were surrounded by an extra-dune environment where aeolian and non-aeolian deposits were inter-mixed. Fryberger *et al.* (1979) and Andrews (1981) documented fluvial channels and playas inter-bedded with aeolian deposits at the Great Sand Dunes. Fryberger (1979) and Langford (1989) described mud-draped aeolian dunes formed through flooding of inter-dune and sand sheet areas. Langford (1989) described fluvial reshaping of aeolian landscapes that resulted in distinctive flooded inter-dune deposits. Breed *et al.* (1979) described numerous examples of sabkhas, streams, playas and sand sheets at the margins of many of the great ergs of the world.

Summary

These descriptions of modern ergs suggest that ancient erg margin deposits can be characterized by one or more of the six aforementioned criteria in the following ways.

1 For upwind margins with a coarse-grained source area, sandstones should be coarser and less well sorted. Downwind margins should be finer-grained and better sorted.

2 Dune set heights should be thinner due to smaller, closer spaced and sometimes more rapidly migrating dunes.

3 Aeolian dune deposits should reflect deposition from simpler and closer spaced dune forms.

4 More abundant, low-angle aeolian deposits should form from deposition in extensive sand sheets, zibars, siliciclastic sabkhas and inter-dunes. Distinctive zibar, sabkha and sand sheet deposits should be recognizable. Inter-dune deposits should be more abundant due to the proportionally greater inter-dune area, but perhaps thinner due to more rapid dune migration.

5 Bioturbation (by roots and burrows) should be present in either aeolian dune or inter-dune sandstone.

6 Intercalated non-aeolian deposits such as playas, fluvial channel fills and sabkhas should be more abundant.

Ancient studies

Studies of ancient strata have interpreted erg margin and erg interior environments based on the same six criteria as those discussed for modern studies.

Grain size and sorting

Coarse-granule horizons and ventifacts were associated with Triassic erg margin deposits in Nova Scotia (Hubert & Mertz, 1980). Clemmensen & Abrahamsen (1983) and Brookfield (1989) noted abundant granule horizons within inferred inter-dunes in Permian erg margin deposits in Scotland. Chan (1989) and Steele (1987) noted abundant coarse-grained laminae in sheet sandstone facies in the White Rim Sandstone in Utah. Kocurek (1981b) described coarse-grained deposits at the fluvial margin of the Entrada erg.

Dune size and dune forms (as inferred from preserved cross-strata)

An upwind erg margin position has been inferred for domal dunes in the Triassic Keuper Sandstone (Thompson, 1969). In the Permian of Scotland, simple barchan and transverse dune forms were interpreted as erg margin dunes (Clemmensen & Abrahamsen, 1983). Permian dune interior deposits were characterized by thick, complex transverse dunes and draas, with only isolated simple barchans. These sandstones were recently reinterpreted as deposits of oblique dunes (Sneh, 1988). In the Rotliegendes, Brookfield (1989) noted the contrast between the simple dunes described from Permian erg margin deposits by Clemmensen & Abrahamsen (1983) and erg interior, complex dunes in the Dumfries basin.

The margins of a Permian erg in the Schnebly Hill Formation in northern Arizona were formed by small blowout dunes, in contrast to the large complex draas in the erg interior (Blakey & Middleton, 1983). Rapidly moving transverse dunes were inferred to form the upwind margin of the Proterozoic Bigbear erg in Canada, whereas complex barchans formed within the interior (Ross, 1983). Thinner foreset coset thicknesses were used to infer an erg margin setting in which dunes migrated more rapidly (Ross, 1983). Smaller parabolic dunes formed the downwind margin of the Quaternary Nebraska sand hills (Ahlbrandt *et al.*, 1983). Porter (1987) described isolated pods of small, simple aeolian dunes from downwind erg margin deposits of the Jurassic Aztec Sandstone in Nevada and southern California. Aeolian dunes were interpreted to become larger and more complex towards the Navajo erg interior (Porter, 1987). Within the Jurassic Entrada Sandstone there were dramatic differences between marine or fluvial margin and erg interior cross-set thicknesses (Kocurek, 1981b). Foreset dip directions were also different at the marine margin. Coastal dunes were interpreted as simple crescentic dunes. Erg interior dunes were much larger, compound crescentic draas. However, significant differences in aeolian dune form were not evident within the dune deposits of the Permian White Rim Sandstone (Chan, 1989). Dune cross-bed sets were thinner within inferred erg margin deposits in Scotland (Clemmensen & Abrahamsen, 1983). Thinner dune set thicknesses were primary evidence in inferring an erg margin origin for portions of the Jurassic Aztec Sandstone (Porter, 1987). Cross-strata set thickness increased from 3–8 m at the erg margin to 5–10 m in the interior. Thin cross-bed sets forming from smaller, simple barchans were interpreted as margin deposits of a Devonian erg in Ireland in contrast to thick sets deposited by complex draas in the erg interior (Carruthers, 1987).

However, different dune types may coexist within erg interiors. Kerr & Dott (1986) and Kerr (1989) describe complex transverse dunes within one portion of the Tensleep erg and much larger complex oblique forms in an adjacent area.

Abundance of sand sheets, zibars and inter-dunes

In the Jurassic Moenave Formation, Clemmensen *et al.* (1989) indicated changes from aeolian dune to sand sheet facies within 1–5 km distances at the erg margin. The sand sheets extended over 30–60 km distances before grading into inland sabkha facies. Widespread sand sheet deposits were described from Permian erg margin deposits in Scotland (Clemmensen & Abrahamsen, 1983). Porter (1987) described small isolated dunes within a matrix of zibars and sand sheets from one margin of the Jurassic Navajo erg. The downwind margin of longitudinal dunes in the Permian Yellow Sands of England was inferred to be a series of sand patches and sheets (Steele, 1983). An extensive sand sheet with isolated small dunes formed on the upwind margin of the Proterozoic Bigbear erg (Ross, 1983). An extensive sand sheet containing siliciclastic sabkha deposits and isolated small simple barchan dunes formed the downwind margin of the Permian White Rim erg (Chan, 1989). An erg margin origin is inferred for a sand sheet facies within the Triassic Dolores Formation in southwestern Colorado (Blodgett, 1984; Kocurek & Nielson, 1986). The facies is characterized by a horizontally bedded sandstone with abundant mud drapes and features associated with wind ripples. Isolated sand sheets have been described from the Entrada Sandstone (Kocurek, 1981b).

Thicker inter-dunal deposits have been used to identify erg margins in Permian sandstones in Scotland, as contrasted to the thicker and more abundant inter-dune deposits inferred to be erg margin deposits (Clemmensen & Abrahamsen, 1983). Brookfield (1989) noted the importance of both tectonically derived geomorphology and wind direction in determining the abundance and extent of inter-bedding of non-aeolian facies. Alternatively, Kocurek (1981a) found that inter-dune deposits were largely absent from erg margin deposits and were most common within the erg interior associated with regularly spaced dunes. Within the erg margin deposits, marine deposits replaced inter-dunes.

Abundance of bioturbation

Ribbon-like trace fossils in the Permian Caspar Sandstone in Wyoming were ascribed to an erg margin environment (Hanley *et al.*, 1971). Well preserved horizontal and vertical trace fossils are found in erg margin facies of the Jurassic Navajo Sandstone (Middleton & Blakey, 1983). Sand sheet deposits in the Triassic Dolores Formation, marginal to the Wingate erg, contain calcretes and abundant root structures and burrows (Blodgett, 1984; Kocurek & Nielson, 1986). Trace fossils including marine traces were interpreted to indicate an erg margin position within the Permian White Rim Sandstone (Kamola & Chan, 1988).

Abundance of non-aeolian deposits

Dunes flanked with fluvially derived mudstone, inter-bedded fluvial facies and aeolian sandstones and rapid lateral transitions from aeolian to fluvial facies characterize both the Jurassic Moenave Formation (Clemmensen *et al.*, 1989) and the Permian Cedar Mesa Sandstone (Langford & Chan, 1989). Mud-filled inter-dunes occur in marine margin deposits of the Entrada Sandstone (Kocurek, 1981b). Along the fluvial margin of the Entrada there was a laterally well defined change from fluvial wadi sandstones to aeolian sandstones. Playa muds and fluvial-channel sandstones are associated with inferred erg margins in the Permian sandstones of Scotland (Clemmensen & Abrahamsen, 1983). Volcaniclastic sediments and alluvial fan, wadi (Marzolf, 1983; Porter, 1987) and meandering and braided fluvial-channel deposits (Marzolf, 1983; Middleton & Blakey, 1983) were identified with marginal deposits of the Jurassic Navajo erg. Marine-derived shales and sabkha deposits are more common at the margins of the Jurassic Entrada erg (Kocurek, 1981b). Sabkha, coastal lagoon and marine limestones are inter-bedded with erg margin aeolian sandstones in the Permian Schnebly Hill Formation (Blakey & Middleton, 1983). Extensive siliciclastic sabkha deposits formed in marine margin inter-dunes in the Minnelusa Formation (Fryberger, 1986). Fluvial arkoses and shales are inter-bedded with aeolian sands at the downwind margin of the Pennsylvanian Weber Sandstone in northern Colorado and the Galtymore Formation in Ireland (Fryberger, 1979; Carruthers, 1987). However, clay-rich, adhesion-rippled aeolian sandstones and associated sabkha deposits were ascribed to erg interior sabkhas in the Permian Rotliegendes (Glennie *et al.*, 1978).

In contrast, some workers have emphasized the ephemeral nature of dune seas (see Kocurek, 1988). Kerr & Dott (1986) and Kerr (1989) have described inter-bedded erg interior deposits and marine mudstones, and Loope (1985) depicts inter-bedded erg interior and marine sandstones, juxtaposed through sea level changes and repeated deflation of the ergs.

Other features used to infer erg margin origins include abundant deflationary erosion surfaces (Hubert & Mertz, 1980; Clemmensen & Abrahamsen, 1983). Langford & Chan (1988) mapped an increase from four erosion surfaces to seventeen at the erg margin within the Permian Cedar Mesa Sandstone in southern Utah. Numerous thin, erosionally bounded upward drying cycles within the Jurassic Moenave Formation on the Colorado Plateau indicated instability within the erg that resulted in fluvial influx and erg expansion and contraction (Clemmensen *et al.*, 1989). Kocurek (1981b) and Kamola & Chan (1988) suggested that large deformation and brecciation features were indicative of a marine erg margin setting.

CEDAR MESA SANDSTONE

The Permian Cedar Mesa Sandstone is well exposed along the Colorado River in the Canyonlands of southeastern Utah. It is a cliff-forming aeolian sandstone, 100–300 m thick (Blakey *et al.*, 1989). The Cedar Mesa Sandstone outcrop is almost continuous along the crest of the Monument Upwarp (Fig. 1). At the northern end of the upwarp in Canyonlands National Park, the Cedar Mesa Sandstone intertongues with fluvial sandstones of the Cutler Formation and marine sandstones and limestones of the lower Cutler beds (Fig. 2). Along its southeast margin, the Cedar Mesa inter-tongues with fluvial and lacustrine deposits of the Cutler Formation. To the west, in the subsurface, the Cedar Mesa intertongues with marine limestones and sandstones

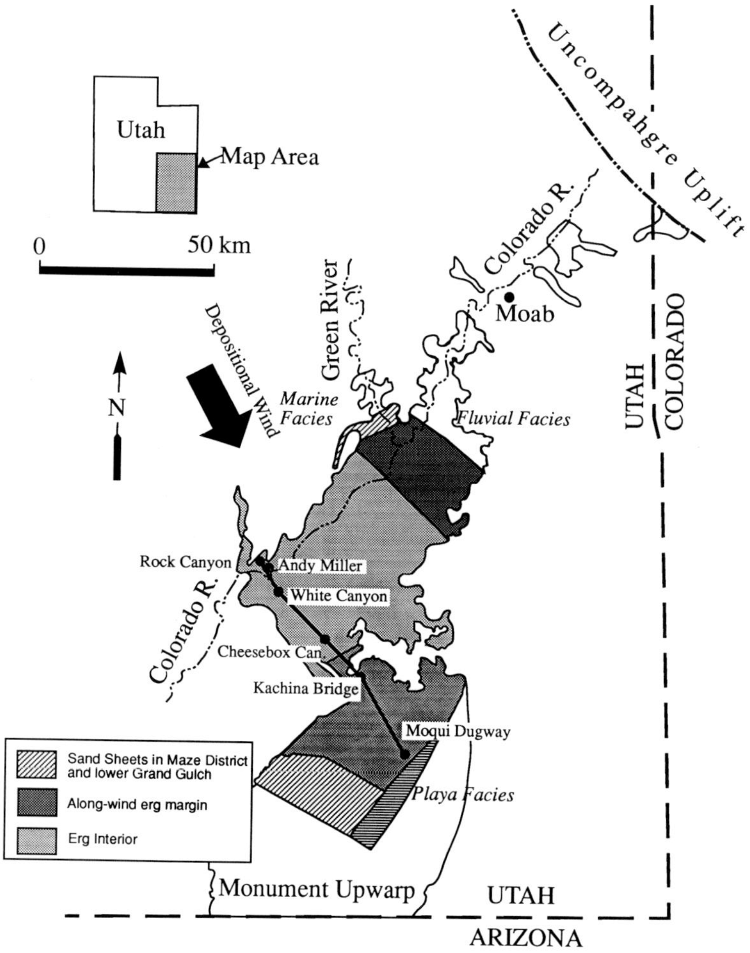

Fig. 1. Map of Cutler Formation and Cedar Mesa Sandstone outcrops and palaeogeographic map of the Cutler–Cedar Mesa depositional system, with outline of the Palaeozoic Uncompahgre Uplift. The arrow indicates wind dispersal directions of aeolian sediment.

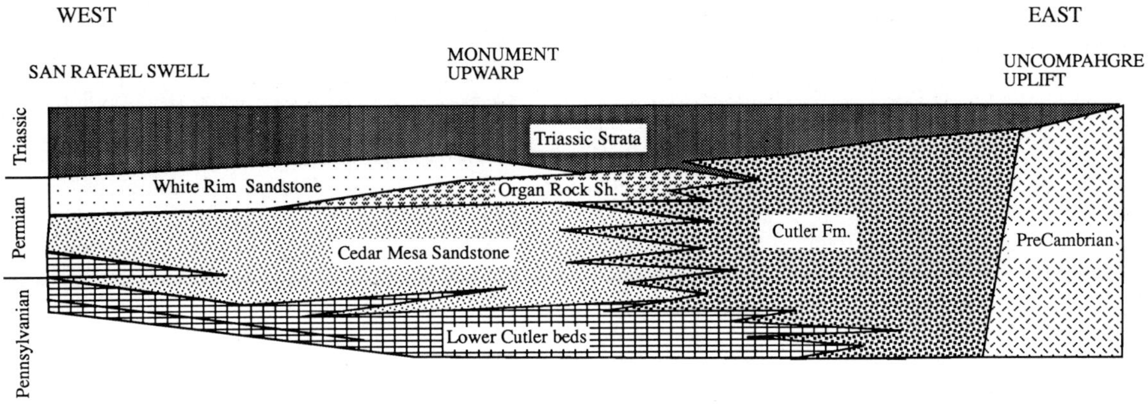

Fig. 2. Schematic east–west cross-section illustrating Permian stratigraphic relations of the Cedar Mesa dune sea.

(Blakey *et al.*, 1989). The dune-forming winds originated in the northwest as inferred from foreset dip directions (Fig. 1) (Mack, 1979; Loope, 1984).

A unique characteristic of the Cedar Mesa Sandstone is that portions of the upwind, downwind and along-wind erg margins are well exposed. The margins of the dune sea were fairly stable during much of the Cedar Mesa deposition and, as a result, thick erg margin aeolian sandstones were deposited within relatively narrow zones (Fig. 1). Exposed zones of inter-tonguing aeolian and non-aeolian deposits occupy well defined bands.

In a previous study of the northeastern (along-wind) margin of the dune sea we observed little change in aeolian deposition from the erg margin into the interior (Langford & Chan, 1988, 1989). We did not quantify changes in grain size or bioturbation, but there were no observable changes in dune types or abundance. The major changes were in the abundance of fluvially deposited inter-beds, and in the abundance of erosion surfaces which decreased from seventeen to four over 18 km.

We have also observed, but not studied, more obvious changes in two areas of the Cedar Mesa outcrop. In the northern part of the Maze District of the Canyonlands, sand sheet deposits replace dunes within a narrow band at the only exposure of the upwind erg margin. In Lower Grand Gulch near the southern margin of the erg, there is a broader band of sand sheets with rare interspersed dunes.

Facies

Four aeolian facies are prevalent within the Cedar Mesa Sandstone. These include three facies distinguished by different aeolian cross-stratification and interpreted as different dune forms or morphologies, and a fourth sand sheet facies.

Large trough sets

This facies is composed of cross-stratified sandstones with foresets 1–7 m thick. Foresets are concave in bedding plane exposures with preserved scours averaging 2–2.4 m thick and 6 m across (Fig. 3A). Basal scours either are concave scoops or have flat bases with concave outer margins. Large troughs may be isolated, but typically occur in clusters of laterally adjacent and superimposed sets. The foresets commonly have low-angle toes (<15°) within the lowest 0.5–1 m. Internally the toes contain thin wind-ripple laminae produced from migrating wind ripples (Hunter, 1977). These ripple-stratified laminae inter-tongue with lenticular, generally massive tongues of sand deposited from avalanches on the dune foresets (Hunter, 1977). Along the foresets, thin laminated regions are irregularly inter-bedded with thicker regions of avalanche deposits. The avalanche tongues are coarsest at the base and become finer-grained and less well defined near the tops of the foresets. The medium-grained avalanche tongues are among the coarsest-grained deposits of the Cedar Mesa Sandstone.

In most bedding plane exposures, the foresets are symmetrically distributed above the basal scours and migration was parallel to the long axes of the scours. Over areas of several square kilometres the scour-trough axes and foreset dips are well aligned, with low dispersion. Trough axes in the same area and stratigraphic unit are usually aligned within 10°.

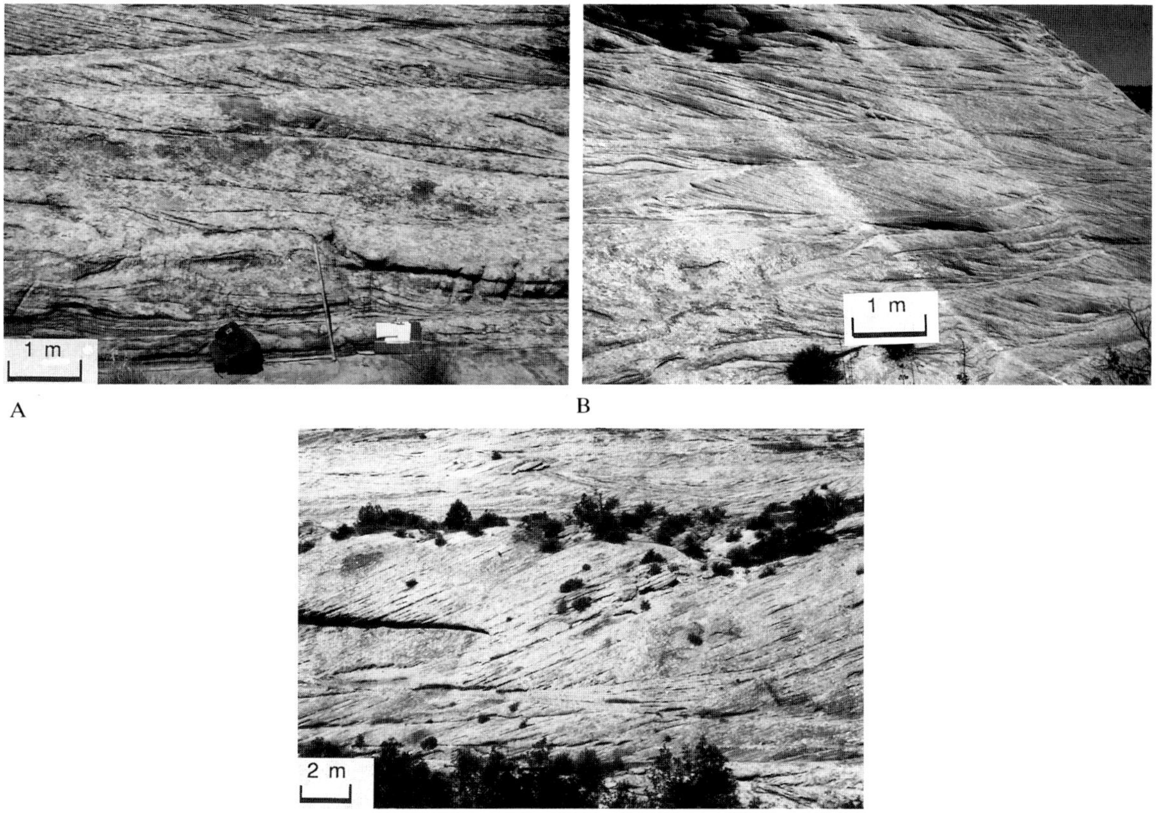

Fig. 3. Photographs of aeolian morphologies in the Cedar Mesa Sandstone. (A) Large-trough cross-stratified sets. (B) Small cross-stratified sets. (C) Planar cross-stratified sets. Note the wedge of low-angle cross-strata formed at the toe of a reactivation surface within the planar cross-set. The relationship results from reshaping (lowering the slope angle) of the foresets.

However, in the same unit at different locations trough axis orientation may vary by as much as 40°. There is also a gradual change from a 155° orientation at Rock and Andy Miller Canyons to a 140° orientation at Kachina Bridge. Large troughs are interpreted as deposits of migrating, three-dimensional bedforms of the crescentic or barchanoid type.

Small trough sets

The small-trough cross-stratified sets are significantly smaller than the large-trough sets, forming a statistically significantly smaller and finer-grained population, averaging 0.6–0.8 m in height (Figs 3B & 4). Like the large-trough sets, the small-trough sets have concave erosional bases and concave foresets. Typical small troughs average 0.5–1 m thick, with preserved scours <3 m across. The internal structures of the small troughs are different from those

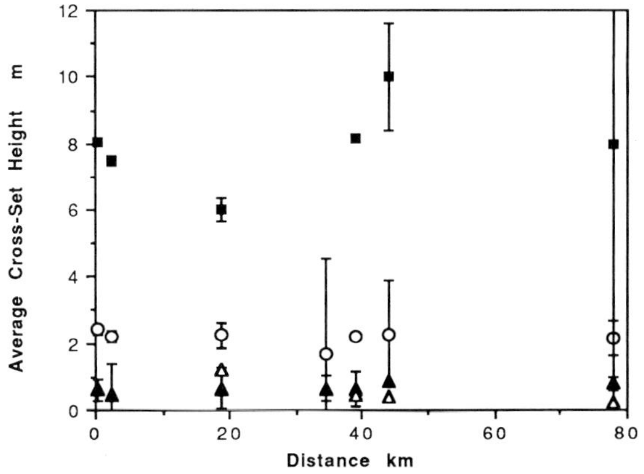

Fig. 4. Average set heights of different dune morphologies: (○) large troughs; (▲) small troughs; (■) planar sets; (△) sand flat. Error bars are 95% confidence intervals.

of the large troughs, consisting predominantly of migrating wind-ripple laminae. Foreset dips are typically lower (<15°) than in large troughs. Small troughs commonly occur in stacked sets separated by subparallel erosional bounding surfaces. These surfaces dip in various orientations but most commonly dip parallel to the axes of the troughs. This implies that the small-trough sets were created by small, three-dimensional dunes migrating down the slope of a larger, low-relief bedform (see Rubin, 1987, fig. 65, for an example). The dips of set boundaries range from 2 to 5°.

Scour-trough axes and foresets dip in the same direction as those of associated large-trough sets. As in the large-trough sets, the foresets are symmetrically distributed within the scours in plan view. The dispersion of well exposed small-trough axes was only slightly higher than of associated large-trough axes.

Planar sets

Planar sets are the thickest cross-stratified sets, averaging 6–8 m. The foresets are similar in cross-section to those of large troughs but are straight and not concave in plan view (Fig. 3C). Extensive bedding plane exposures reveal that the foresets are linear and extend in some instances for at least 400 m. The foresets dip perpendicularly to the axes of associated small and large troughs.

Avalanche deposits form the greatest proportion of the planar sets, becoming finer-grained upward within the foresets. Steeply inclined avalanche tongues may reach the base of the foreset, or the foreset may terminate in a basal wind-ripple dominated 'toe', similar to those of the troughs. Some foresets intertongue upward into low-angle beds composed of flat-bedded wind-ripple laminae and small-trough sets. Reactivation surfaces are very common and stratal relationships indicate that this was due to reshaping (lowering the slope angle) of the foresets (Fig. 3C). The structures produced by changing bedform asymmetry are unique (Rubin, 1987) and (e.g. Fig. 3C) probably resulted from large-scale reshaping of the dune by wind directions differing from those of the foreset-producing winds.

The planar sets are also notably coarse-grained and are as coarse as or coarser than associated large troughs (Fig. 5). Planar cosets almost invariably occur as isolated sets, surrounded by other facies. Set heights are thicker than those of other associated sets and do not decrease downwind (Fig. 4).

Planar sets are interpreted as deposits of migrating, two-dimensional dune forms that migrated transversely to the predominant wind direction. The reactivation surfaces indicate that the dunes were commonly reshaped by winds at high angles to the predominant depositional winds.

Sand sheet deposits

Sand sheet facies are thin-bedded deposits consisting of low-angle (<4°) laminae enclosing numerous thin cross-sets of the small-trough type. Sand sheets are the finest-grained of the aeolian facies (Fig. 5). Beds are lenticular, separated by low-angle erosion surfaces. Bioturbation by plants and animals commonly obscures sedimentary structures. Sedimentary structures are poorly defined, and grain size is relatively homogeneous vertically through sand sheet units. Typical small-trough sets average 0.5–1 m thick, with preserved scours <3 m across; excepting their discontinuous occurrence, they are similar to

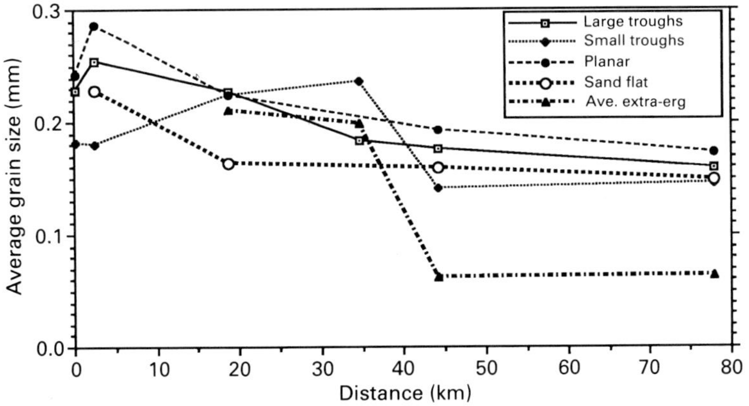

Fig. 5. Average grain size of aeolian facies plotted against distance downwind.

the small troughs described above. Troughs may be abundant and sand sheets may grade laterally into the small-trough facies.

TRANSECT

For this study, erosion surfaces in the uppermost Cedar Mesa Sandstone were traced along an 80 km transect (Figs 1 & 6). Although the Cedar Mesa is not incised deeply enough to expose the erosion surfaces along approximately 15 km of the transect, the relative continuity and extent of the exposed portions of the surfaces suggest that they are continuous beneath the covered intervals. Correlation of deflation surfaces allows identification of relatively synchronous deposits within the Cedar Mesa Sandstone.

Methods

Six stratigraphic sections were measured along the transect (Figs 1 & 6–11). Bioturbation was measured on a semi-quantitative scale, created by Droser & Bottjer (1986), and converted to a percentage. The percentage of a 0.25 m^2 area that was bioturbated or in which sedimentary structures were obscured was measured. The thickness of bioturbated intervals was recorded. Thicknesses of low-angle aeolian deposits and non-aeolian deposits were measured as a proportion of measured stratigraphic sections. Additional measurements included dune set heights and the thickness of inferred inter-dune beds.

The distribution of sedimentary structures was mapped for a 1 m wide zone along the measured sections. Cross-bed and bounding surface orientations were measured and used to obtain the dune morphology by the methods proposed by Rubin & Hunter (1983). Grain size was measured by visually estimating the average grain size at 1 m intervals within each stratigraphic section. Where a coarser population of grains was present, its average grain size and the percentage of total grains were estimated. Samples taken from fifty of these measurement sites were thin-sectioned. Two hundred grain diameters in each sample were measured under

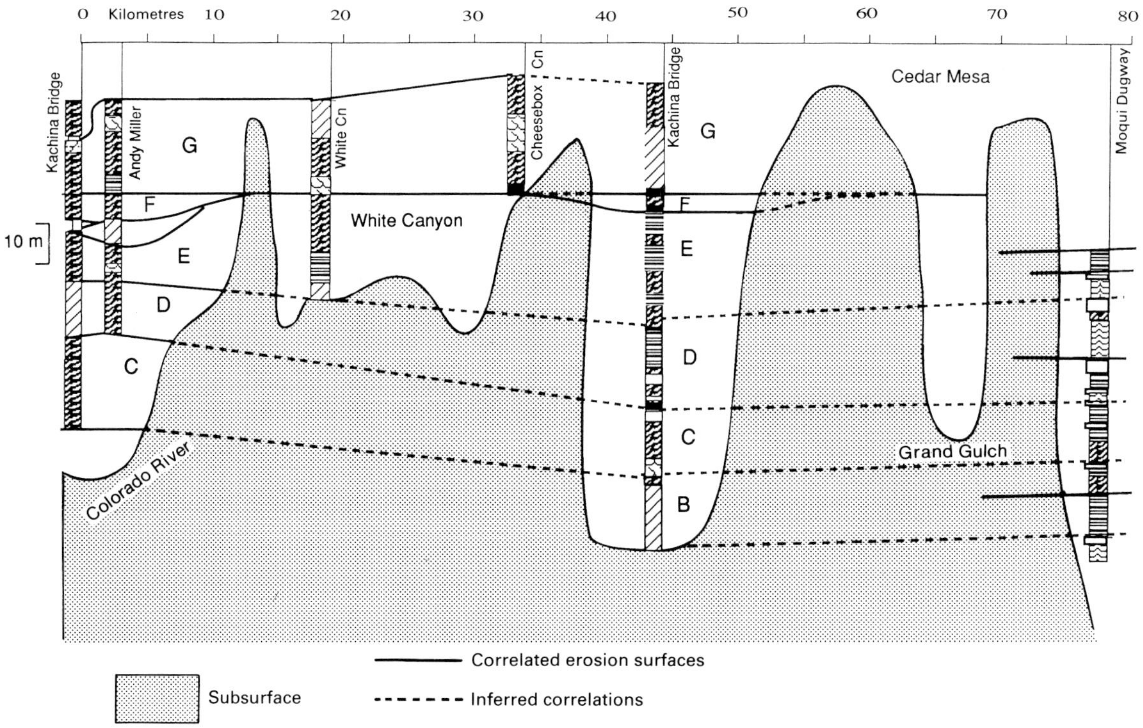

Fig. 6. Cross-section along transect showing the location of measured sections and exposure along the transect.

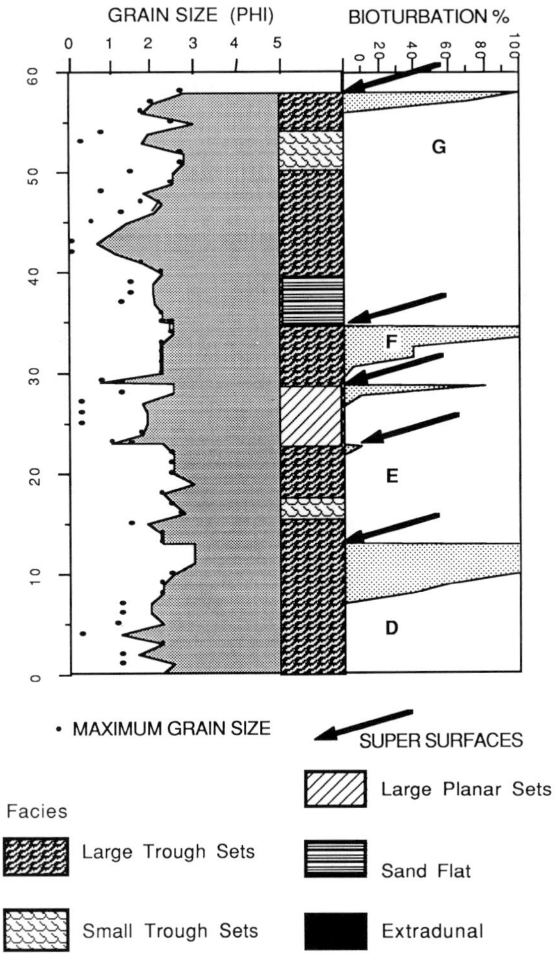

Fig. 7. Andy Miller section.

a polarizing microscope to calibrate the visual measurements.

INTERNAL STRATIGRAPHY

Supersurfaces

Extensive erosion surfaces stratigraphically subdivide the Cedar Mesa dune sea (Loope, 1985; Langford & Chan, 1988). These 'supersurfaces' are commonly underlain by patches of bioturbated aeolian sandstone and are irregular, with up to 15 m of relief over a 3 km distance. Locally, several metres of relief are evident. These have been interpreted as surfaces of aeolian erosion because of associated subaerial fossils and trace fossils and association

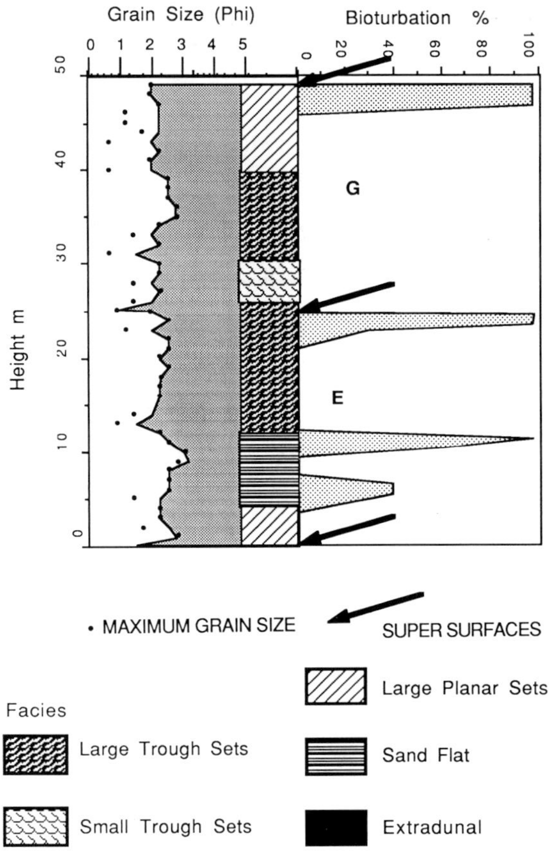

Fig. 8. White Canyon section.

with aeolian depositional features (Loope, 1984, 1985).

The more extensive surfaces represent widespread deflationary events that can be traced over 100–1000 km^2 areas within the Cedar Mesa outcrop belt (Loope, 1985). Some of these extend across the entire transect and are used here to define the internal stratigraphy of the formation (Fig. 6). These are labelled as supersurfaces separating stratigraphic units A–G (Figs 7–11). The identification of the supersurfaces that extend from Moqui Dugway to Kachina Bridge sections is uncertain. Throughout the rest of the transect, surfaces were walked out for correlation purposes. The lower surfaces are not exposed in upper White Canyon between the White Canyon and Kachina Bridge and thus correlation here is inferred. These extensive supersurfaces are probably deflation surfaces on which net erosion occurred across the dunefield. Loope (1985) inferred

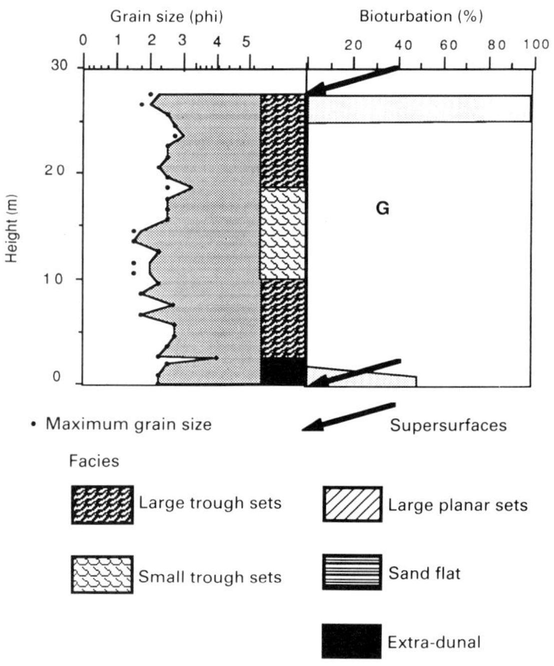

Fig. 9. Cheesebox Canyon section.

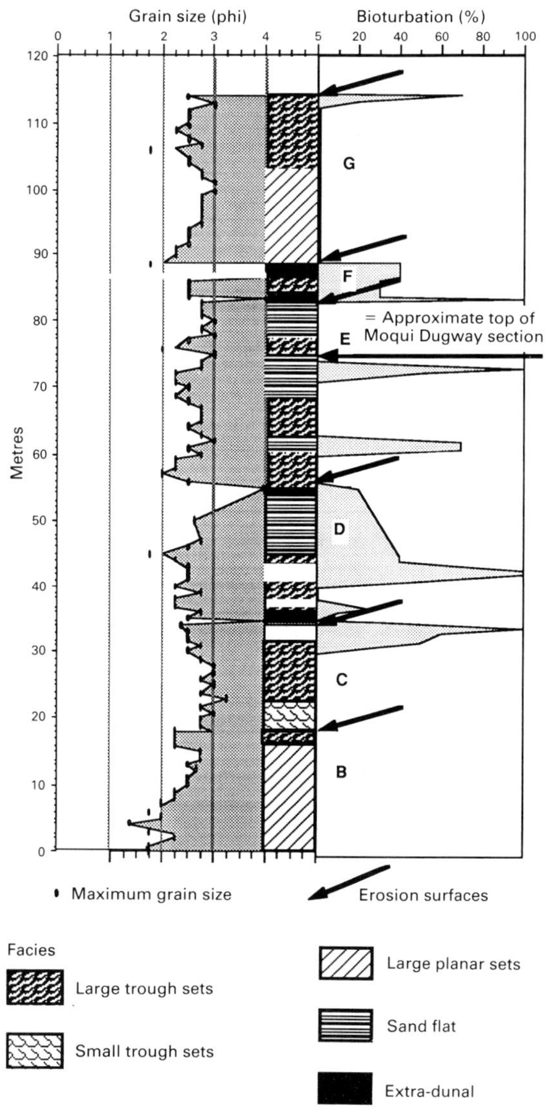

Fig. 10. Kachina Bridge section.

that aeolian deposition in the Cedar Mesa was episodic and the deflation surfaces were formed during regional erosional events. In this study, this type of surface will be referred to as a *deflation surface* following the use by Loope (1985).

Less extensive erosion surfaces are associated with the margins of the dune sea and occur between the deflationary surfaces (Langford & Chan, 1988). These surfaces represent more localized aeolian erosion and are associated with non-aeolian (fluvial) deposits. Coexistence of non-aeolian and aeolian deposition may be demonstrated, indicating that the less extensive surfaces formed during periods of erosion within an actively aggrading erg (Langford & Chan, 1988, 1989). These supersurfaces will be referred to as *local supersurfaces*, although they also mark at least localized and short-term deflation. Langford & Chan (1988) named these 'flood surfaces'.

DOWNWIND CHANGES IN DEPOSITION

Grain size

The Cedar Mesa Sandstone is noticeably coarser-grained than most of the Palaeozoic and Mesozoic aeolian sandstones of the Western Interior. This happy circumstance results in easily measurable grain size changes at all scales. Grain size varied along the transect due to a number of factors. There was a well defined down-wind decrease in average grain diameters of approximately 0.16 mm km^{-1} (Fig. 12). The average of the diameters of the coarsest fraction decreased approximately 0.27 mm km^{-1} (Fig. 12). The average of the coarser fractions decreased rapidly to Kachina Bridge, 44 km downwind, and

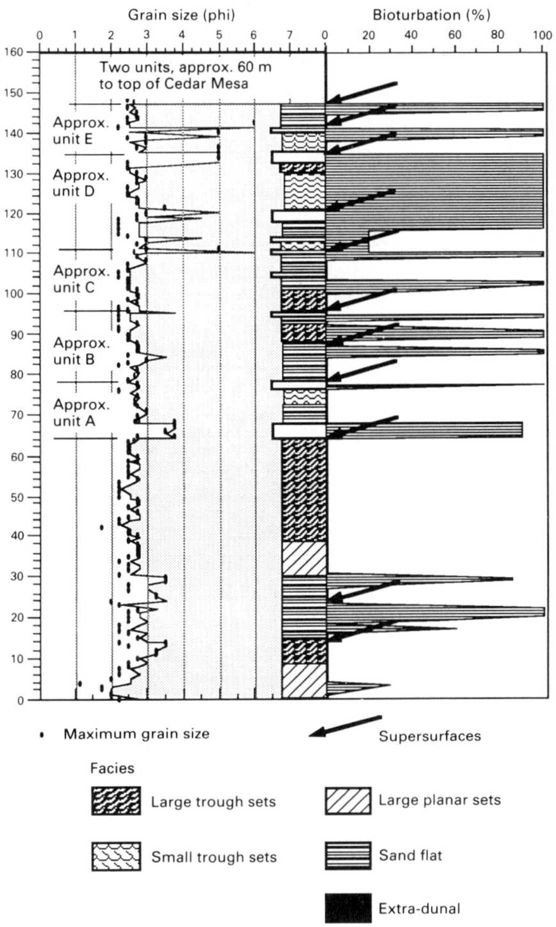

Fig. 11. Moqui Dugway section.

thereafter decreased more slowly. The histograms of average grain sizes indicate a bimodal population in the upwind sections. Therefore, a chi-square test was run for all of the sections and for pairs of sections. The chi-square variable increased with distance between sections (Fig. 12). It was significant at a 95% level of confidence for sections more than 20 km apart.

Grain size also varies between facies (Fig. 5). The coarsest facies in the upwind sections are large troughs and planar types. The average grain size of these facies was significantly coarser (at a 95% level of confidence) than that of small troughs and bioturbated horizons at Rock Canyon and Andy Miller Canyon (Fig. 1). At White Canyon, the mean grain sizes for large troughs, small troughs and planar sets were significantly coarser than the mean for sand sheets. At Kachina Bridge and Moqui Dugway, small troughs and sand sheets are finer-grained than large troughs and planar deposits but are much coarser-grained than fluvial deposits.

Although all of the aeolian facies decrease in grain size downwind, a proportion of the downwind decrease in grain size is accounted for by a downwind decrease in the abundance of large troughs, from 75% in the interior of the dune sea to 30% near the downwind margin.

Individual cross-sets were finer-grained higher up the foresets. Grain size was found to decrease slightly, corresponding with foreset height on thick planar and trough sets (Fig. 13). The lower foresets were predominantly grainflow deposits, with about 70% grainflow and 30% grainfall deposits nearer the top. The lowest 1 m of these dunes was commonly a

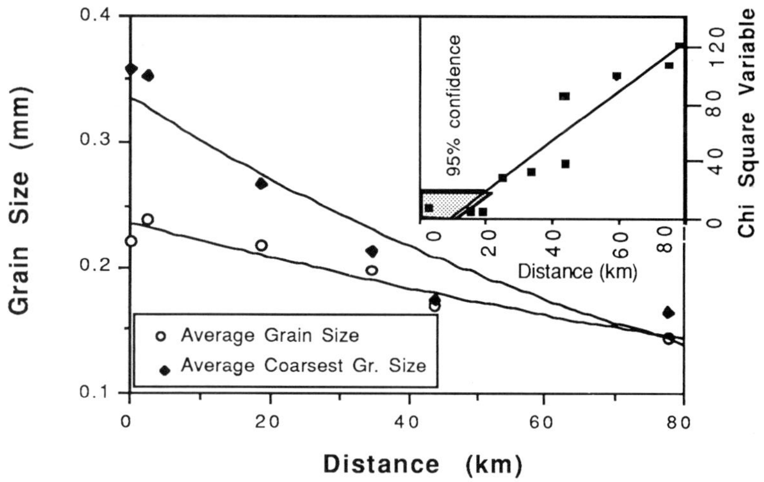

Fig. 12. Average mean and average maximum grain size plotted against distance downwind.

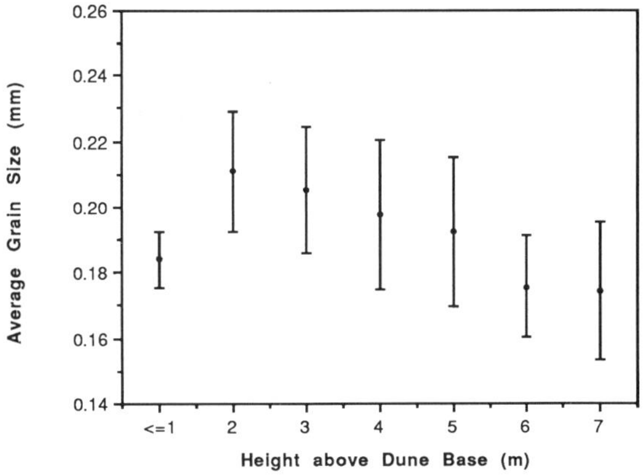

Fig. 13. Mean grain size as a function of height above the base of thicker cross-stratified sets. Error bars are 95% confidence intervals.

wind-ripple laminated 'foreset toe' and was also finer-grained. A plot of grain sizes against height above the base of the dunes reveals that a more restricted grain size is found at the tops of the foreset, the range decreasing from 0.55 mm at the base of dunes to 0.04 mm at the top.

The uppermost unit (G) within the Cedar Mesa was consistently coarser-grained than the immediately underlying unit (E) (Fig. 14). The difference in average and average maximum grain size would correspond approximately to a shift of 10 km in the position of the source of aeolian sand (Fig. 14).

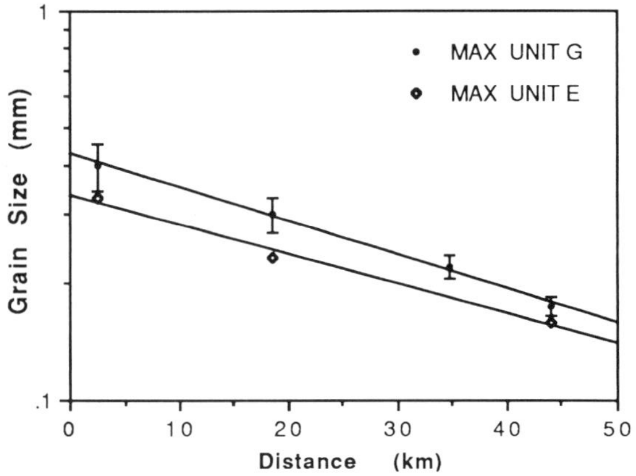

Fig. 14. Grain size plotted against distance for units E and G, the uppermost continuous units in the Cedar Mesa Sandstone. Note the consistently coarser mean for the uppermost unit, G. Error bars are 95% confidence intervals. (See Figs 6–10 for stratigraphic correlation.)

However, the uppermost unit is less bioturbated, and because bioturbated intervals tend to be composed of fine- and very fine-grained sandstone, the difference in grain size may result from differential preservation of vegetated areas.

Bioturbated horizons

Bioturbated horizons are intervals within which sedimentary structures are largely indistinguishable due to biological activity. Bioturbated horizons commonly underlie both deflation and local supersurfaces. Some sand sheets and fluvial deposits are also extensively bioturbated. In the upwind sections (Rock Canyon, Andy Miller Canyon and White Canyon) the bioturbated horizons are the finest-grained intervals. In the remaining downwind sections the grain size is near the average for all facies. The Moqui Dugway section, within 10 km of the downwind margin of the Cedar Mesa Sandstone, contains a significantly higher percentage of bioturbated sandstone (Fig. 11). However, the percentage of bioturbated intervals within the other five sections offers no obvious trend, varying widely around a value of approximately 15% of the section.

The percentage of bioturbated sandstones is obviously partially related to the abundances of non-aeolian and sand sheet facies within the measured sections. Although the percentage of bioturbated dune sandstone decreases downwind, the percentage of bioturbated sand sheet and non-aeolian deposits increases (Fig. 15). Bioturbation also varies stratigraphically. Twelve per cent of the uppermost stratigraphic interval is bioturbated, whereas 29% of the underlying unit is bioturbated (Figs 7, 8 & 10). This variation is independent of distance from the edge of the erg and abundance of non-aeolian deposits. Independence of stratigraphic or non-aeolian influences suggests changes in climate or groundwater table fluctuations as possible causes of bioturbation.

Dune set height

The average preserved set heights for the different dune types do not change significantly downwind along the transect (Fig. 4). The heights of preserved cross-stratified sets are controlled by many factors. Cross-strata are preserved through the climb of migrating bedforms relative to the depositional surface and through the migration of scours (Rubin & Hunter, 1982, 1983). The thickness of cross-stratified

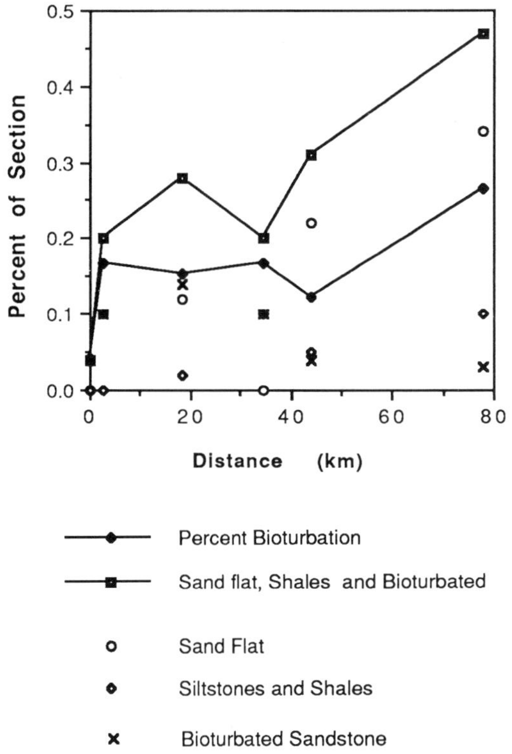

Fig. 15. Plot of bioturbation as a percentage of the stratigraphic section versus distance downwind. For comparison, the percentages of non-aeolian and sand sheet facies are also plotted.

sets created by bedform climb is related to the spacing, speed of migration and angle of climb of the dunes. The depositional surface may be a foreset itself and inclined to the underlying and overlying supersurfaces. The heights of foresets filling migrating scours are controlled by the same consideration in addition to the dimensions and rate of migration and climb of the scours and associated features (see Rubin, 1987, for examples). Regardless of the contributing factors, changes in dune size, spacing, rate of migration or rate of climb will result in variation in preserved cross-stratal dimensions (Kocurek, 1981; Rubin & Hunter, 1982; Kerr, 1989).

Large- and small-trough cross-stratified sets are created by downwind migration of scour pits or three-dimensional dunes. The consistent set thickness requires that the depth of scour pits or dunes remained fairly constant and that the amount of climb between successive scour pits or dunes remained fairly constant along the transect.

Planar cross-stratified sets occur as isolated sets. On some of the dunes, flat-lying top-set beds composed of small-trough cross-sets and sand sheet deposits are preserved. Preservation of top sets and of planar sets of consistent thickness indicates that these sets represent a fundamental bedform height. Because the average bedform heights remain consistent downwind, the size of bedforms probably did not change downwind.

Local supersurfaces

Another indicator of the proximity of the margins of the dune sea are the supersurfaces within the aeolian sandstones (Figs 7–11). Numerous additional local supersurfaces appear within 20 km of the erg margins (Fig. 11), increasing in frequency at the erg margin from one per 17 m of section in the five upwind sections to one per 9 m of section at Moqui Dugway. The additional local supersurfaces are commonly overlain by aqueously deposited siltstones and shales and underlain by bioturbated aeolian sandstone.

Both local and deflation supersurfaces are interpreted as marking periods of aeolian deflation associated with decreases in the aeolian sand supply. Wilson (1973) suggested that ergs are localized by saturation of the aeolian transporting winds with sand. Aggradation of ergs occurs when winds are sand-saturated. Deflation occurs when winds are undersaturated with sand. Wilson (1973) mapped sand coverage in Saharan ergs and inferred that ergs are characterized by central regions in which winds are supersaturated and aggradation occurs. Erg margins are characterized by saturation of transporting winds where sand is transported as migrating aeolian dunes but aggradation does not occur. Loope (1990) suggests that erg margins are characterized by more frequent periods of undersaturation, resulting in the formation of additional local supersurfaces.

SUMMARY AND CONCLUSIONS

Analysis of the transect allows several inferences to be made about the Cedar Mesa dune seas that deposited its uppermost stratigraphic units.

1 The well defined downwind decrease in grain size and the consistent average grain size at each section indicate that the sources of aeolian sand did not change either in position or in sand composition throughout the periods of deposition. Had the grain size of transportable sand changed through time, the grain size within the resultant aeolian sandstones should fluctuate.

There are two end-member combinations of wind regime and sandflow that result in accumulation of sand. The first type of erg is that in which sand derived from distant or local sources is deposited as the effectiveness of the wind regime to transport sand is gradually reduced. These ergs are characterized by wide zones along the upwind margins across which sand is transported, with deposition concentrated near the erg margins towards which sand is transported. In this case, the prime controls are the large-scale wind regime and geomorphology that concentrate transported sand. Wilson (1973) described a model for this type of erg using the Saharan dune seas as examples.

In the second type of erg, aeolian sand immediately saturates the transporting wind and is deposited near the source. Further from the sand source, the winds are not saturated and sand flowing off the downwind margins of these ergs is transported away from the system. These areas are characterized by sand deposition near the source, with a long 'tail' of non-depositional, transporting dunes at the downwind margin. The primary control on this type of dune sea is the sand supply and the wind regime at that point. The outcropping part of the Cedar Mesa Sandstone was probably deposited by this type of dune sea, tied to a coastal sand source. Numerous marine fossil fragments in the upwind aeolian deposits (Mack, 1978, 1979) and the abundance of coarse- and very coarse-grained sand in upwind sections of the Cedar Mesa support this interpretation.

The second type of erg may gradually migrate further downwind from its source as slowly moving, coarse aeolian debris mantles areas adjacent to the source and the zone of sand saturation is shifted downwind. Bagnold (1941) has described this type of dune sea in the Great Western erg. However, assuming that the sediment supply is constant, migrating ergs should be recorded as upward-coarsening deposits. The Cedar Mesa maintains consistent grain sizes at all localities throughout its deposition (Figs 7, 10 & 11).

2 Individual deflation surface bounded stratigraphic units do not change in grain size upward. This has implications for the nature of the erosional periods that created the extensive deflation surfaces. If significant aeolian transportation occurred during long periods of erosion, then the coarse-grained material composing the upwind sandstones should have been reworked downwind as a coarse-grained lag. Since this reworking has not occurred, then either (i) the periods of erosion were not long-lasting, or (ii) the sand supply was shut off during deflationary periods either due to removal of the supply or through a change in the transporting wind direction.

3 Within the Cedar Mesa Sandstone:

(a) bioturbation appears to vary stratigraphically more than spatially perhaps due to climatic fluctuations;

(b) dune spacing and morphologies of individual dune types do not vary downwind;

(c) grain size decreases gradually downwind;

(d) proportions of different dune types change regularly as the abundance of large-trough sets decreases downwind;

(e) the proportion of sand sheet deposits increases regularly downwind;

(f) non-aeolian facies increase towards the erg margin but form a significant proportion of the Cedar Mesa at Kachina Bridge, 30–40 km into the dune sea;

(g) supersurfaces increase in frequency at the erg margin from one per 17 m of section in the five upwind sections to one per 9 m of section at Moqui Dugway.

To what extent do changes in depositional features within aeolian sandstones indicate expansion, contraction or migration of the erg? The conclusions reached above indicate that the downwind margin of the Cedar Mesa erg did not expand and contract and that periods of erosion that produced the deflationary supersurfaces were not long-lived. The best evidence for this are the well defined grain size trends. Likewise, changes in the frequency of erosion surfaces and non-aeolian sediment would provide qualitative information about migration or erg expansion. However, changes in the thickness or kind of aeolian depositional facies and bioturbation would not be useful for understanding the history of the erg.

The Cedar Mesa example is our attempt to evaluate a variety of criteria that have been cited from other modern and ancient examples. The most significant conclusion is that the criteria that best demonstrate marginality are different within different ergs and along different margins of the same erg. Along its upwind and its southern margin (Fig. 1), the Cedar Mesa is transitional to sand sheet deposits.

REFERENCES

AHLBRANDT, T.S., ANDREWS, S. & GWYNNE, D.T. (1978) Bioturbation in eolian deposits. *J. Sed. Petrol.* **48**, 839–848.

AHLBRANDT, T.S. & FRYBERGER, S.G. (1981) Sedimentary features and significance of interdune deposits. In: *Recent and Ancient Nonmarine Depositional Models for Exploration* (Ed. Etheridge, F.G. & Flores, R.M.) pp. 293–314. Spec. Publ. Soc. Econ. Paleont. Mineral., Tulsa, 31.

AHLBRANDT, T.S., SWINEHART, J.B. & MARONEY, D.G. (1983) The dynamic Holocene dune fields of the Great Plains and Rocky Mountain Basins, USA. In: *Eolian Sediments and Processes* (Ed. Brookfield, M.E. & Ahlbrandt, T.S.) pp. 379–406. Dev. Sediment. 38. Elsevier, Amsterdam.

ANDREWS, S. (1981) Sedimentology of Great Sand Dunes, Colorado. In: *Recent and Ancient Nonmarine Depositional Models for Exploration* (Ed. Etheridge, F.G. & Flores, R.M.) pp. 293–314. Spec. Publ. Soc. Econ. Paleont. Mineral., Tulsa, 31.

ASH, J.E. & WASSON, R.J. (1983) Vegetation and sand mobility in the Australian desert dunefield. *Z. Geomorph. NF Suppl-bd* **45**, 7–25.

BAGNOLD, R. (1941) *The Physics of Blown Sands and Desert Dunes*. Chapman & Hall, London, 265 pp.

BLAKEY, R.C. & MIDDLETON, L.T. (1983) Permian shoreline eolian complex in central Arizona: dune changes in response to cyclic sea level changes. In: *Eolian Sediments and Processes* (Ed. Brookfield, M.E. & Ahlbrandt, T.S.) pp. 551–582. Dev. Sediment. 38. Elsevier, Amsterdam.

BLAKEY, R.C., PETERSON, F. & KOCUREK, G. (1989) Synthesis of late Paleozoic and Mesozoic eolian deposits of the western interior of the United States. *Sed. Geol.* **56**, 3–125.

BLODGETT, R.H. (1984) Nonmarine depositional environments and paleosol development in the Upper Triassic Dolores Formation, southwestern Colorado. *Guidebook, 37th Annual Meeting, Rocky Mountain Section of the Geological Society of America, Durango*, 46–92.

BREED, C.S., FRYBERGER, S.G., ANDREWS, S., MCCAULEY, C., LENNARTZ, F., GEBEL, D. & HORSTMAN, K. (1979) Regional studies of sand seas using Landsat (ERTS) imagery. In: *A Study of Global Sand Seas* (Ed. McKee, E.D.) pp. 305–398. US Geol. Surv. Prof. Pap. 1052.

BREED, C.S. & GROW, T. (1979) Morphology and distribution of dunes in sand seas observed by remote sensing. In: *A Study of Global Sand Seas* (Ed. McKee, E.D.) pp. 253–304. US Geol. Surv. Prof. Pap. 1052.

BROOKFIELD, M.E. (1989) Rotliegendes desert sedimentation in Scotland. *Z. Geol. Wiss.* **17**, 205–241.

CARRUTHERS, R.A. (1987) Aeolian sedimentation from the Galtymore Formation (Devonian), Ireland. In: *Desert Sediments: Ancient and Modern* (Ed. Frostick, L.E. & Reid, I.) pp. 251–268. Spec. Publ. Geol. Soc. London 35.

CHAN, M.A. (1989) Erg margin of the Permian White Rim Sandstone, SE Utah. *Sedimentology* **36**, 235–251.

CLEMMENSEN, L.B. & ABRAHAMSEN, K. (1983) Aeolian stratification and facies association in desert sediments, Arran basin (Permian) Scotland. *Sedimentology* **25**, 587–604.

CLEMMENSEN, L.B., OLSEN, H. & BLAKEY, R.C. (1989) Erg-margin deposits in the lower Jurassic Moenave Formation and the Wingate Sandstone, southern Utah. *Bull. Geol. Soc. Am.* **101**, 759–773.

DROSER, M. & BOTTJER, D.J. (1986) A semi-quantitative field classification of ichnofabric. *Geol. Soc. Am., Cordilleran Section 82nd Annual Meeting, Abstract with Programme 18*, 102.

FOLK, R.L. (1971) Longitudinal dunes of the northwestern edge of the Simpson desert, Northern Territory, Australia. 1. Geomorphology and grain-size relationships. *Sedimentology* **16**, 5–54.

FRYBERGER, S.G. (1979) Eolian–fluviatile (continental) origin of ancient stratigraphic trap for petroleum in Weber Sandstone, Rangely oil field, Colorado. *Mount. Geol.* **16**, 1–36.

FRYBERGER, S.G. (1986) Stratigraphic traps for petroleum in wind-laid rocks. *Bull. Am. Ass. Petrol. Geol.* **70**, 1765–1776.

FRYBERGER, S.G., AHLBRANDT, T.S. & ANDREWS, S. (1979) Origin, sedimentary features, and significance of low-angle 'sand sheet' deposits, Great Sand Dunes National Monument and vicinity, Colorado. *J. Sed. Petrol.* **49**, 733–746.

FRYBERGER, S.G., AL-SARI, A.M. & CLISHAM, T.J. (1983) Eolian dune, interdune, sand sheet, and siliciclastic sabkha sediments of an offshore prograding sand sea, Dhahran area, Saudi Arabia. *Bull. Am. Ass. Petrol. Geol.* **67**, 280–312.

GLENNIE, K.W. (1970) Desert sedimentary environments. *Dev. Sed.* **14**, 222 pp.

GLENNIE, K.W., MUDD, G.C. & NAGTEGAAL, P.J.C. (1978) Depositional environment and diagenesis of Permian Rotliegendes sandstones in the Lehman Bank and Sole Pit areas of the UK southern North Sea. *J. Geol. Soc. Lond.* **135**, 25–34.

HANLEY, J.H., STEIDTMANN, J.R. & TOOTS, H. (1971) Trace fossils from the Casper Sandstone (Permian) southern Laramie Basin, Wyoming and Colorado. *J. Sed. Petrol.* **41**, 1065–1068.

HUBERT, J.F. & MERTZ, K.A. (1980) Eolian dune field of Late Triassic age, Fundy Basin, Nova Scotia. *Geology* **8**, 516–519.

HUNTER, R.E. (1977) Basic types of stratification in small eolian dunes. *Sedimentology* **24**, 361–387.

KAMOLA, D.L. & CHAN, M.A. (1988) Coastal dune facies, Permian Cutler Formation (White Rim Sandstone), Capitol Reef National Park area, southern Utah. *Sed. Geol.* **56**, 341–356.

KERR, D.R. (1989) *Sedimentology and stratigraphy of Pennsylvanian and lower Permian Strata (Upper Amsden Formation and Tensleep Sandstone) in North Central Wyoming*. Unpublished PhD thesis, University of Wisconsin.

KERR, D.R. & DOTT, R.H. (1986) Eolian dune types preserved in the Tensleep Sandstone (Pennsylvanian–Permian), north-central Wyoming.

KOCUREK, G. (1981a) Significance of interdune deposits and bounding surfaces in eolian dune sands. *Sedimentology* **28**, 753–780.

KOCUREK, G. (1981b) Erg reconstruction: the Entrada Sandstone (Jurassic) of northwestern Utah and Colorado. *Palaeogeog., Palaeoclim., Palaeoecol.* **36**, 125–153.

KOCUREK, G. (1988) First-order and super bounding

surfaces in eolian sequences—bounding surfaces revisited. *Sed. Geol.* **56**, 193–206.

KOCUREK, G. & NIELSON, J. (1986) Conditions favorable for the formation of warm-climate aeolian sand sheets. *Sedimentology* **33**, 795–816.

LANCASTER, N. (1982) Dunes of the skeleton coast, Namibia (South West Africa): geomorphology and grain size relationships. *Earth Surf. Proc. Landf.* **7**, 575–587.

LANCASTER, N. (1983) Controls of dune morphology in the Namib sand sea. In: *Eolian Sediments and Processes* (Ed. Brookfield, M.E. & Ahlbrandt, T.S.) pp. 261–290. Dev. Sediment. 38. Elsevier, Amsterdam.

LANCASTER, N. (1986) Grain-size characteristics of linear dunes in the southwestern Kalihari. *J. Sed. Petrol.* **56**, 395–400.

LANCASTER, N. (1988) The development of large aeolian bedforms. *Sed. Geol.* **55**, 69–89.

LANGFORD, R.P. (1989) Modern and ancient fluvial–eolian interactions. Part 1, Modern systems. *Sedimentology* **36**, 1023–1035.

LANGFORD, R.P. & CHAN, M.A. (1988) Flood surfaces and deflation surfaces within the Cutler Formation and Cedar Mesa Sandstone (Permian), southeastern Utah. *Bull. Geol. Soc. Am.* **100**, 1541–1549.

LANGFORD, R.P. & CHAN, M.A. (1989) Modern and ancient fluvial–eolian interactions. Part 2, Ancient systems. *Sedimentology* **36**, 1037–1051.

LOOPE, D.B. (1984) Eolian origin of upper Paleozoic sandstones, southeastern Utah. *J. Sed. Petrol.* **54**, 563–580.

LOOPE, D.B. (1985) Episodic deposition of upper Paleozoic Sandstones, southeastern Utah. *Geology* **13**, 73–76.

LOOPE, D.B. (1990) Personal communication.

MACK, G.H. (1978) The survivability of light-mineral grains in fluvial, eolian and marine environments. The Permian Cutler and Cedar Mesa Formations. *Sedimentology* **25**, 587–604.

MACK, G.H. (1979) Littoral marine depositional model for the Cedar Mesa Sandstone (Permian), Canyonlands National Park. In: *Permianland* (Ed. Baars, D.L.) pp. 53–68. Four Corners Geol. Soc. Ninth Field Conference, Durango.

MAINGUET, M. & CHEMIN, M.C. (1983) Sand seas on the Sahara and the Sahel: an explanation of their thickness and sand dune type by the sand budget principle. In: *Eolian Sediments and Processes* (Ed. Brookfield, M.E. & Ahlbrandt, T.S.) pp. 353–364. Dev. Sediment. 38. Elsevier, Amsterdam.

MARZOLF, J.E. (1983) Changing wind and hydrologic regimes during deposition of the Navajo and Aztec Sandstone, Jurassic (?), southwestern United States. In: *Eolian Sediments and Processes* (Ed. Brookfield, M.E. & Ahlbrandt, T.S.) pp. 635–660. Dev. Sediment. 38. Elsevier, Amsterdam.

MCKEE, E.D. (1966) Structures of dunes at White Sands National Monument, New Mexico (and a comparison with structures of dunes from other selected areas). *Sedimentology* **7**, 1–69.

MIDDLETON, L.T. & BLAKEY, R.C. (1983) Processes and controls on the intertonguing of the Kayenta and Navajo Formations, northern Arizona: eolian–fluvial interactions. In: *Eolian Sediments and Processes* (Ed. Brookfield, M.E. & Ahlbrandt, T.S.) pp. 613–634. Dev. Sediment. 38. Elsevier, Amsterdam.

NIELSON, J. & KOCUREK, G. (1986) Climbing zibars of the Algodones. *Sed. Geol.* **48**, 1–15.

PORTER, M.L. (1987) Sedimentology of an ancient erg margin: the Lower Jurassic Aztec Sandstone, southern Nevada and southern California. *Sedimentology* **34**, 661–680.

ROSS, G.M. (1983) Bigbear erg: a Proterozoic intermontane eolian sand sea in the Hornby Bay Group, Northwest Territories, Canada. In: *Eolian Sediments and Processes* (Ed. Brookfield, M.E. & Ahlbrandt, T.S.) pp. 483–520. Dev. Sediment. 38. Elsevier, Amsterdam.

RUBIN, D.M. (1987) Cross-bedding, bedforms, and paleocurrents. *Soc. Econ. Paleontol. Mineral., Concepts in Sed. and Paleontol. 1*. SEPM, Tulsa, 187 pp.

RUBIN, D.M. & HUNTER, R.E. (1982) Bedform climbing in theory and nature. *Sedimentology* **29**, 121–138.

RUBIN, D.M. & HUNTER, R.E. (1983) Reconstructing bedform assemblages from compound cross-bedding. In: *Eolian Sediments and Processes* (Ed. Brookfield, M.E. & Ahlbrandt, T.S.) pp. 407–428. Dev. Sediment. 38. Elsevier, Amsterdam.

SNEH, A. (1988) Permian dune patterns in northwestern Europe challenged. *J. Sed. Petrol.* **58**, 44–51.

STEELE, R.P. (1983) Longitudinal draa in the Permian Yellow Sands of north-east England. In: *Eolian Sediments and Processes* (Ed. Brookfield, M.E. & Ahlbrandt, T.S.) pp. 543–550. Dev. Sediment. 38. Elsevier, Amsterdam.

STEELE, B.R. (1987) Depositional environments of the White Rim Sandstone member of the Cutler Formation, Canyonlands National Park Utah. *Bull. US Geol. Surv. 1592-B*, 20 pp.

TALBOT, M.R. (1985) Major bounding surfaces in aeolian sandstones, a climatic model. *Sedimentology* **32**, 257–265.

THOMPSON, D.B. (1969) Dome-shaped eolian dunes in the Frodsham Member of the so-called 'Keuper' Sandstone Formation (Scythian–Frasnian: Triassic) at Frodsham, Cheshire (England). *Sed. Geol.* **3**, 263–289.

WASSON, R.J. (1983) Dune sediment type, sand colour, sediment provenance and hydrology in the Strzelecki-Simpson dunefield, Australia. In: *Eolian Sediments and Processes* (Ed. Brookfield, M.E. & Ahlbrandt, T.S.) pp. 165–196. Dev. Sediment. 38. Elsevier, Amsterdam.

WILSON, I.G. (1973) Ergs. *Sed. Geol.* **10**, 77–110.

Spec. Publs Int. Ass. Sediment. (1993) **16**, 127–149

Low-stand aeolian influence on stratigraphic completeness: upper member of the Hermosa Formation (latest Carboniferous), southeast Utah, USA

S.C. ATCHLEY* *and* D.B. LOOPE
Department of Geology, University of Nebraska, Lincoln, NE 68588, USA

ABSTRACT

The Late Pennsylvanian (latest Carboniferous) record of the Paradox Basin of southeast Utah includes a thick sequence of cyclically inter-bedded marine, fluvial and aeolian deposits. We have reconstructed the areal distribution and relative water depths of marine depositional environments on the basis of lithology and facies stacking order within fourth-order cycles (established through Markov chain analysis). Estimates of water depths for facies make it possible to subdivide fourth-order cycles into transgressive and regressive fifth-order cycle sets. Fourth-order cycles are most often disconformably overlain by non-marine low-stand deposits. The facies associations within fourth-order cycles suggest superimposition on a third-order cycle. Third-order cycles lie within a (second-order) episode of long-term decrease in accommodation space. We attribute the higher-frequency (fifth- and fourth-order) cycles to the waxing and waning of Gondwana glaciers, while the longer-term second- and (perhaps) third-order cycles may reflect variations in tectonic accommodation provided by subsidence.

Allocyclic interpretations generally imply that the observed cycles can be correlated basin-wide. While the second- and third-order cycles can be correlated for a distance of at least 90 km, the fourth- and fifth-order cycles are difficult to correlate, even locally. This anomaly may be explained by the aridity of the Paradox Basin during Late Pennsylvanian time. Due to an arid palaeoclimate and a calcite dominated composition, grain supported marine carbonates remained unlithified after exposure. During fourth-order low stands, some of these sediments were reworked into aeolian dunes. The marine record was partially or totally stripped away, making correlation of fourth- and fifth-order cycles difficult. The fundamental nature of the basin fill was thus strongly influenced by deflation.

INTRODUCTION

Purpose

In the Paradox Basin of southeast Utah, inter-bedded carbonate and siliciclastic strata of the Late Pennsylvanian (latest Carboniferous) upper member of the Hermosa Formation (Fig. 1) record cyclic transitions from marine to terrestrial depositional processes. Hite (1960) and Goldhammer *et al.* (1989) noted cyclicity of a similar scale in the evaporite dominated Middle Pennsylvanian Paradox Member of the Hermosa Formation and other workers have reported cyclic Pennsylvanian strata from other regions and continents (e.g. Wanless & Shepard, 1936; Heckel, 1980). Unique to the west-central USA, however, is that cyclic late Palaeozoic rocks record the systematic superimposition of shelf carbonates by aeolian sands (Driese & Dott, 1984; Loope, 1985; Loope & Haverland, 1988; Rice & Loope, 1991). The aeolian deposits frequently include a high percentage of marine carbonate bioclasts and lithoclasts.

What was the mechanism for cyclicity? It is difficult to imagine an autocyclic process that could (i) juxtapose shelf and aeolian facies, or (ii) lead to the development of deep desiccation fissures (Loope & Haverland, 1988) and oxidation of pyritized fossils and traces within the marine units (Loope & Watkins, 1989). Quaternary aeolian carbonates have been considered a low-latitude signature of glaciation

* Present address: Exxon Production Research Company, PO Box 2189, Houston, TX 77252-2189, USA.

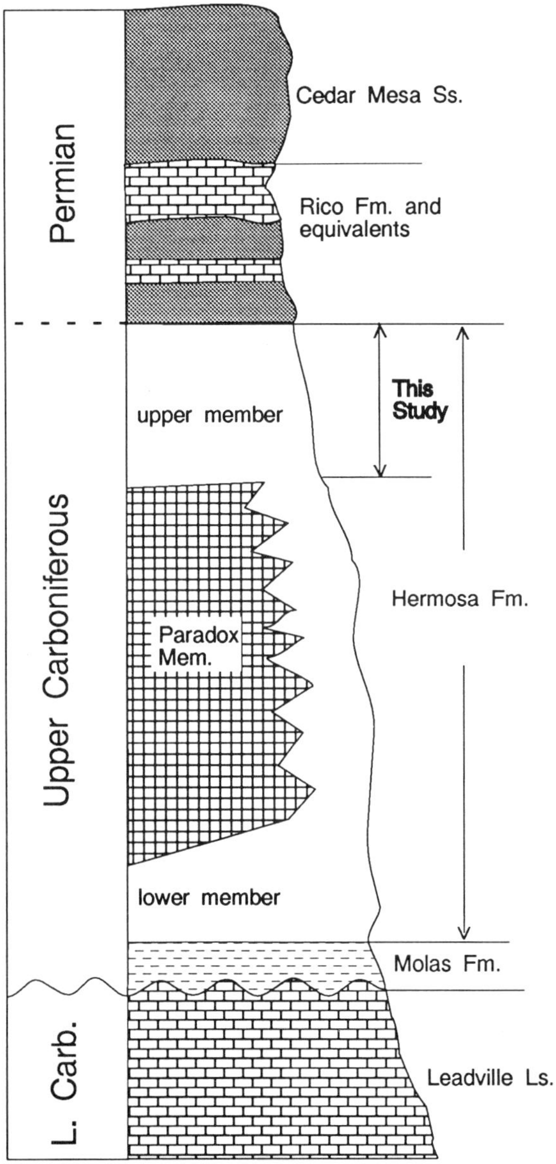

Fig. 1. Generalized stratigraphic column showing relationship of upper member of Hermosa Formation (this study) to the evaporite-bearing Paradox Member of the Hermosa Formation (Goldhammer *et al.*, 1989). (Stratigraphic terminology after Hite & Buckner, 1981.)

(Johnson, 1968; Fairbridge & Johnson, 1978). From palaeomagnetic and sedimentological data, it has been demonstrated that the southern margin of Gondwana passed over the southern pole and was glaciated during the late Palaeozoic (Caputo & Crowell, 1985; Veevers & Powell, 1987). Boardman & Malinky (1985), Heckel (1986) and Goldstein (1988) have argued for glacio-eustatic control of late Palaeozoic North American Mid-continent and Rocky Mountain cyclothems. Furthermore, Boardman & Heckel (1989) have correlated Texas cyclothems one-for-one with the Mid-continent, extending the evidence for an inter-regional allocyclic mechanism for cyclicity.

Clearly, however, subsidence during Middle to Late Pennsylvanian time provided accommodation space for these strata. Could 'jerky' subsidence be the cause of some (or all) of the cyclicity? As pointed out by Hardie (1986) and Koerschner & Read (1989), jerky subsidence alone cannot explain the emergence of marine facies — 'trampoline tectonics' must be invoked for the fourth- and fifth-order cycles.

Can individual upper Hermosa cycles be correlated across the Paradox Basin? Lateral continuity is commonly put forward as the best test of autocyclic vs allocyclic hypotheses (Heckel, 1977; Kozar *et al.*, 1990). Since the vertical sequence of facies and the distribution of diagenetic features in our sections are incompatible with an autocyclic origin, these cycles — whether due to eustasy or to jerky subsidence — should, like other allocycles (e.g. Posamentier *et al.*, 1988; Posamentier & Vail, 1988; Sarg, 1988; Van Wagoner *et al.*, 1988), correlate basin-wide. If, as suggested by Lewis & Campbell (1965), these cycles cannot be correlated basin-wide, perhaps the Late Pennsylvanian marine record of the Paradox Basin is, relative to published examples from other basins, incomplete. Although the lowstand sedimentary record includes both aeolian and fluvial deposits, our observations indicate that aeolian deflation caused this incompleteness. We claim, as did Dalrymple *et al.* (1986) and Rice & Loope (1991), that the fundamental nature of the stratigraphic record of arid (and pre-vegetation) intervals was strongly influenced by deflation.

We have attempted to determine the preferred succession of facies between unconformities, and have used the relative water depths of marine facies to construct a hierarchy of cyclicity. We name our cycles fifth-order (average 9 m thick), fourth-order (average 25 m thick) and third-order (average 213 m thick), based on their similarity in thickness to the fifth-, fourth- and third-order cycles documented by Goldhammer *et al.* (1989) from the immediately underlying Paradox Member of the Hermosa Formation.

Geographic and stratigraphic setting

Five sections were measured within the southern and central portions of the Paradox Basin along canyon walls of the Colorado and San Juan Rivers and their tributaries (Fig. 2). Surface sections were augmented by two subsurface sections (Fig. 2). We will use the stratigraphic nomenclature of Hite & Buckner (1981), and will focus exclusively on the upper member of the Hermosa Formation (Honaker Trail Formation of Wengerd, 1973), which for simplicity will be referred to as the upper Hermosa. The rocks described in this paper represent the penultimate filling of the Paradox Basin, and therefore have distributions limited by the structural elements which surrounded the basin during the latest Pennsylvanian. At that time, the Paradox Basin was an elongate northwest – southeast trough bordered on the northeast by the Uncompahgre Uplift, to the west and northwest by the Emery (Piute) Positive Area, to the southwest by the Kaibab Arch and to the south by the Defiance – Zuni Positive Area (Blakey, 1980) (Fig. 2). The Paradox Basin was actively subsiding during Pennsylvanian time, causing increased deposition towards the basin axis. The upper Hermosa increases in thickness from 250 m along the southwest basin margin (Lewis & Campbell, 1965), to more than 1600 m along the basin axis (Frahme & Vaughn, 1983).

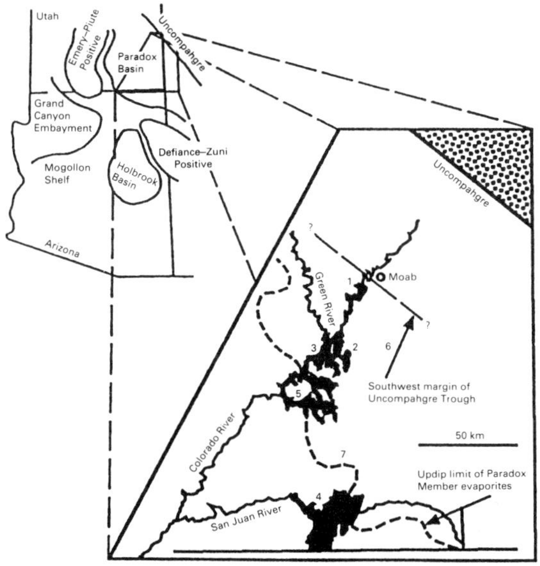

Fig. 2. Study area in the Paradox Basin, southeast Utah. Five surface sections were measured along the canyons of the Colorado and San Juan Rivers: 1, Cane Creek Anticline; 2, Big Spring Canyon; 3, Spanish Bottom; 4, Honaker Trail; and 5, Dark Canyon. Two subsurface cores augmented these surface sections: 6, Gibson Dome No. 1; and 7, Elk Ridge No. 1. Also shown are the structural elements which confined the deposition and therefore extent of marine strata during the Late Carboniferous, and the updip limit of Paradox Member evaporites, which approximates the configuration of the Paradox Basin just prior to upper Hermosa deposition. (Modified from Blakey, 1980; Hite & Buckner, 1981; Loope, 1984.)

FACIES ANALYSIS

Interpretation of the preferred facies succession, and therefore cycles within the upper Hermosa, requires the establishment of criteria that first allow facies differentiation. Outcrop and subsurface sections were constructed with attention to general lithology, component bioclastic and siliciclastic grains, and sedimentary structures. Facies boundaries were categorized as sharp, gradational, covered or missing (in core). Thin sections were prepared from the 259 samples that were collected and a visual estimate of percentage and maximum size of bioclastic and siliciclastic grains was recorded, along with a visual estimate of percentage of carbonate/clay matrix and/or calcite cement. Eleven marine and non-marine lithofacies were identified. Averaged compositional totals for samples from each facies, along with carbonate texture, grain type and characteristic sedimentary structures, allowed the generation of a facies composite (Table 1). Figures 3 & 4 provide photomicrographs representative of each facies.

In making an environmental interpretation for a given facies, Enos (1983) suggests that no single sedimentological or biological element be considered diagnostic. Rather, he recommends consideration of the vertical rock sequence, with emphasis on facies transitions. Walther's Law states that the vertical succession of facies results from the lateral migration of juxtaposed depositional environments. Intuitively, the reconstruction of lateral facies relations is a straightforward task, since these relations are recorded successively in the vertical sequence. A complication, as pointed out by Middleton (1973), is that Walther's Law applies to stratigraphic successions without major breaks. Facies separated by unconformities do not necessarily represent the juxtaposed deposits of contiguous depositional environments.

Table 1. Descriptions and interpretations of facies and their depositional environments; lettering scheme reflects spectrum of environments from deepest marine (A) to non-marine (K)

	Facies designation										
	A	B	C	D	E	F	G	I	J	H	K
Facies interpretation	Euxinic basin	Open marine slope	Algal mound to shoal	Shoal	Hypersaline lagoon	Shoreface to foreshore	Shoreface to foreshore	Coastal plain fluvial and/or loess	Coastal plain fluvial	Aeolian	Lacustrine
Texture and/or classification	Organic-rich mudstone	Wackestone	Boundstone	Packstone to grainstone	Mudstone	Ooid grainstone	Siliciclastic packstone to grainstone	Calcareous argillaceous siltstone	Micaceous arkose to lithic arkose	Siliciclastic sandstone to bioclastic grainstone	Mudstone
Bedding and sedimentary structures	Grain-flow (turbidite) laminae	Massive to inclined laminae	Massive	Planar cross-beds, abundant burrows	Massive to laminated	Planar cross-beds	Planar cross-beds, occasional burrows	Massive to thinly laminated, rhizoliths, mudcracks	Low-angle tabular and small-scale trough cross-beds	Large-scale, high-angle troughs, rhizoliths	Massive to horizontal laminae
Dominant carbonate grains	Crinoid, brachiopod, bryozoan	Bryozoan, crinoid, foram, rugose coral, brachiopod	Phylloid algae	Bryozoan, crinoid, foram, brachiopod	Brachiopod, ostracode	Ooid	Bryozoan, crinoid, foram, brachiopod	Lithoclasts and diverse bioclasts	Lithoclasts and diverse bioclasts	Lithoclasts and diverse bioclasts	Ostracode
Percentage carbonate grains and average maximum size	Trace (−0.50 ϕ) $N = 6$	41% (−1.40 ϕ) $N = 39$	13% (−2.50 ϕ) $N = 3$	29% (−0.64 ϕ) $N = 22$	3% (0.43 ϕ) $N = 19$	53% (0.50 ϕ) $N = 15$	19% (1.60 ϕ) $N = 19$	3% (0.50 ϕ) $N = 31$	10% (1.20 ϕ) $N = 11$	11% (1.87 ϕ) $N = 91$	3% (1.50 ϕ) $N = 3$
Percentage siliciclastics and average maximum size	16% (3.30 ϕ) $N = 6$	8% (2.80 ϕ) $N = 39$	— $N = 3$	26% (2.60 ϕ) $N = 22$	20% (3.30 ϕ) $N = 19$	3% (2.50 ϕ) $N = 16$	40% (2.50 ϕ) $N = 19$	47% (3.10 ϕ) $N = 31$	54% (1.70 ϕ) $N = 11$	45% (2.50 ϕ) $N = 91$	1% (3.50 ϕ) $N = 3$
Inter-granular volume as percentage matrix, cement and/or clay	84% $N = 6$	51% $N = 39$	87% $N = 3$	45% $N = 22$	77% $N = 19$	44% $N = 15$	41% $N = 19$	50% $N = 31$	36% $N = 11$	44% $N = 91$	96% $N = 3$

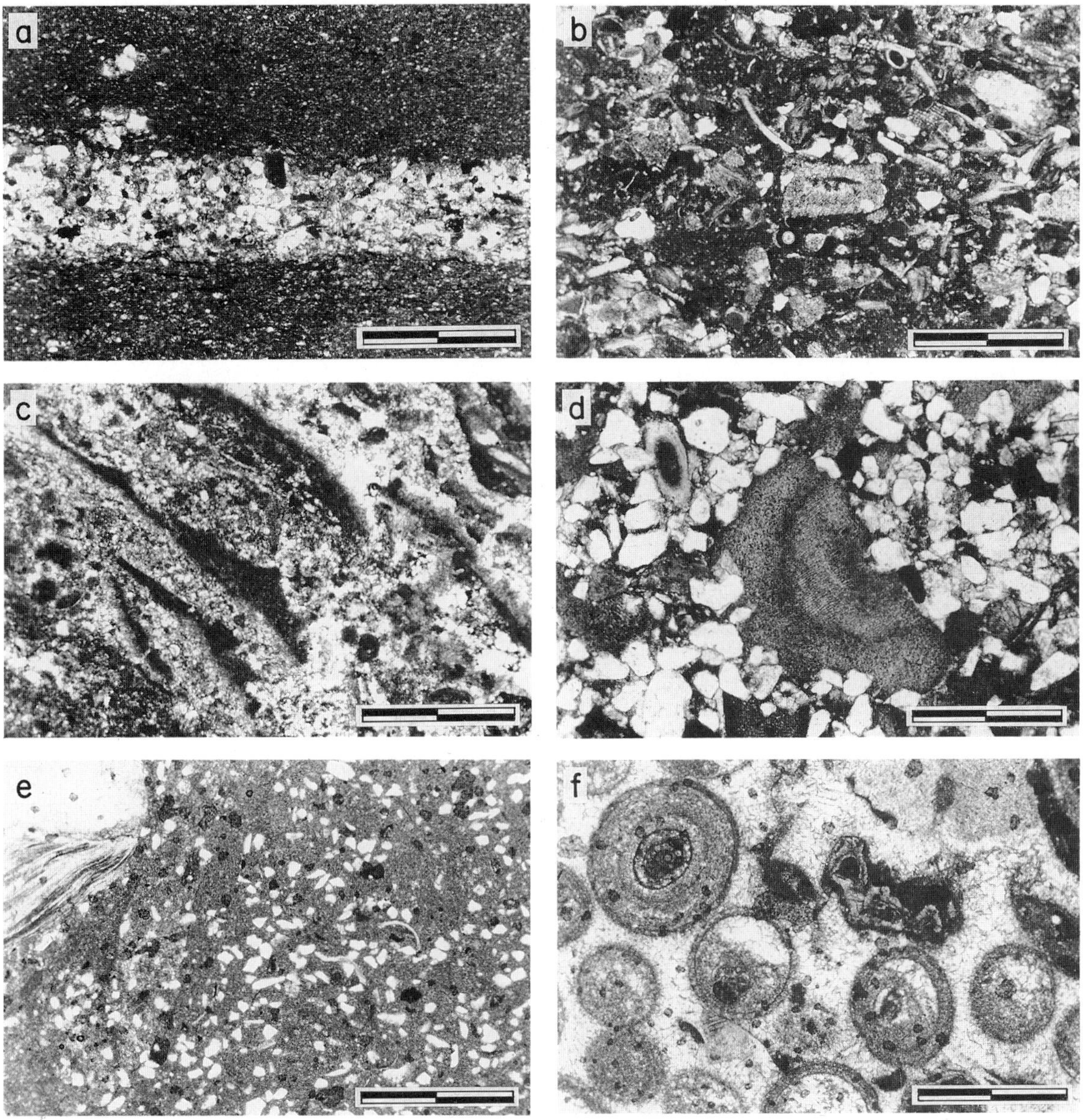

Fig. 3. Photomicrographs representative of facies. Bar scale = 1 mm. (a) Facies A; Gibson Dome core at 759 m. Note the lamina along which relatively coarse siliciclastic grains are concentrated. (b) Elk Ridge core at 661 m. Diagnostic of this facies is the abundance of diverse fossil fragments supported by micrite matrix, and the relatively low percentage of siliciclastic grains. (c) Facies C; Elk Ridge core at 638 m. Diagnostic of this facies are phylloid algal grains observed trending from the upper left to lower right across the photograph. (d) Facies D; Elk Ridge core at 482 m. Note high concentration of relatively coarse siliciclastic and bioclastic grains which have fitted and/or sutured contacts. (e) Facies E; Gibson Dome core at 541 m. Fossil fragments are uncommon. Note large brachiopod fragment (upper left), and disarticulated ostracod valve (middle right). Very fine siliciclastic grains are common, but are suspended within micrite matrix. (f) Facies F; Gibson Dome core at 397 m. The internal structure of ooids is rarely preserved, but when present, shows concentric laminae. Note open fabric.

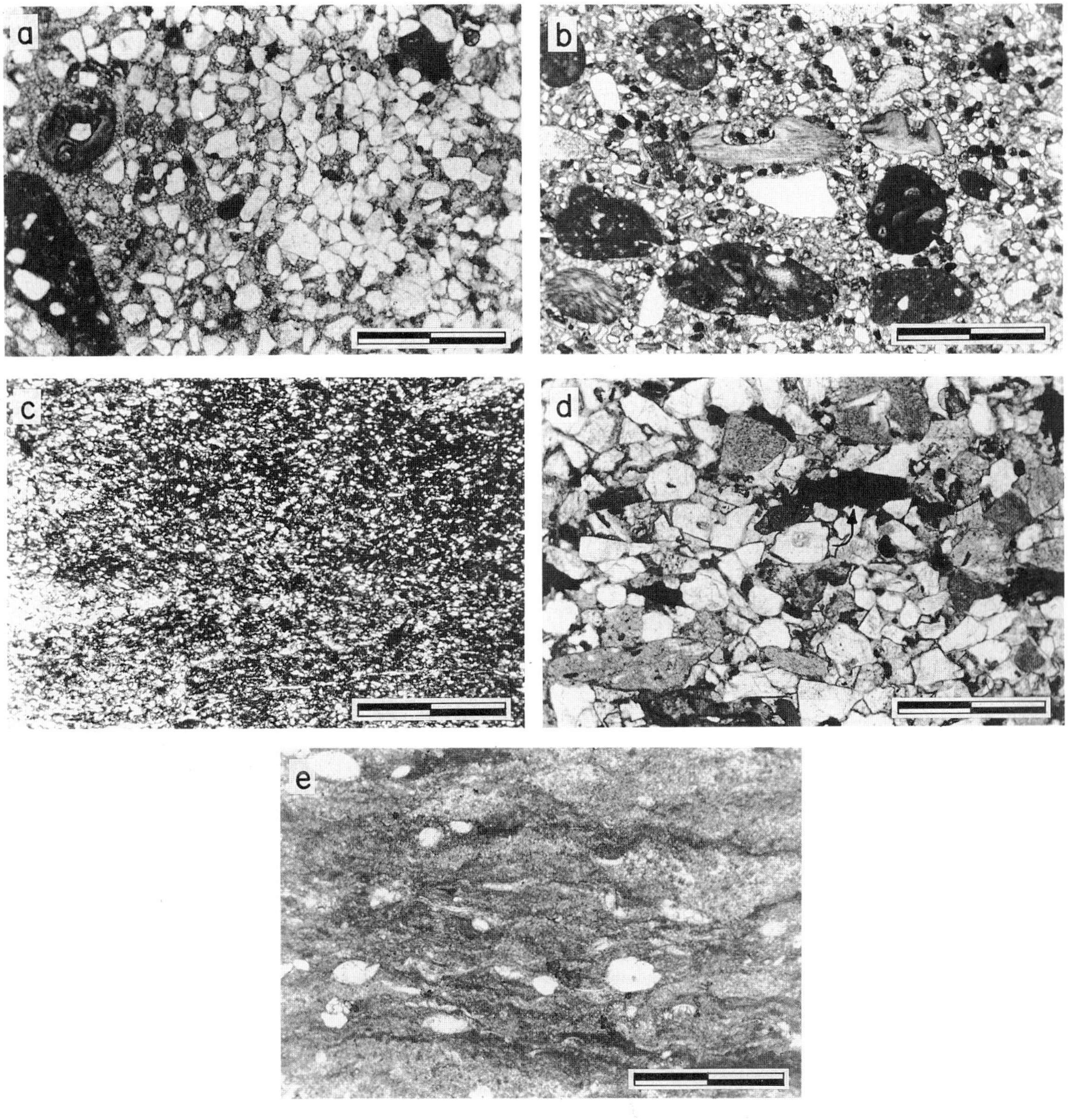

Fig. 4. Photomicrographs representative of depositional facies. Bar scale = 1 mm. (a) Facies G; approximately 165 m above base of Dark Canyon section. (b) Facies H; approximately 38 m above base of Dark Canyon section. Note bimodality of the siliciclastic and highly abraded carbonate grains. Many of the abraded carbonate lithoclasts are reworked fragments from facies B and E. (c) Facies I; Gibson Dome core at 423 m. (d) Facies J; approximately 46 m above base of Cane Creek Anticline section. This facies is typically composed of medium- to coarse-grained, relatively immature siliciclastics. Also diagnostic is the presence of coarse mica grains (arrow). (e) Facies K; approximately 149 m above base of Dark Canyon section. The facies is characteristically a laminated mudstone, with fairly common ostracods (moulds in the photomicrograph). Note lack of siliciclastic grains, even though it most commonly occurs as inter-beds within aeolian facies H.

Most recently, Busch & Rollins (1984), Goodwin & Anderson (1985) and Van Wagoner *et al.* (1988) suggest that the marine stratigraphic record is largely composed of (geologically) instantaneous flooding surfaces (diastems) overlain by progradational sedimentary deposits. The resulting cycles are thin (3–30 m), and record a shoaling succession of gradational environments equal to and shallower than the water depth of the instantaneous base-level rise. Although these workers apply different terms to describe these shoaling upward cycles, we simply describe them qualitatively as fifth-order cycles, based on similarities with those previously documented by Goldhammer *et al.* (1989) in the Paradox Member of the Hermosa, which directly underlies the upper member. Our fifth-order cycles were defined as those gradational successions of marine lithofacies or individual marine lithofacies which are bounded by sharp contacts with vertically adjacent lithofacies. These highest-frequency cycles were most obvious in slabbed core from the Elk Ridge and Gibson Dome sites.

Using the method of Walker (1979), we applied Markov chain analysis to the gradational marine facies successions within fifth-order cycles (Table 2). In the generation of the transition count matrix, a total of 56 row/column mismatches were recorded; we interpret these to reflect the analysis of only unidirectional progradational facies transitions (Table 2). To determine the significance of each gradational facies transition, binomial probabilities were calculated according to the procedure of Harper (1984) (Table 3). The smaller the binomial probability, the more significant the transition. Consideration of these binomial probabilities, therefore, suggests that fifth-order cycles have the following preferred progradational marine facies succession:

Deep A → B → D → E → F → G Shallow

Since facies C was rarely observed on outcrop or in core, its assigned relative water depth is based less on statistical information, and more on the qualitative observation that facies C was only observed gradationally preceded by facies B, and gradationally succeeded by facies E. Consequently, facies C is tentatively considered to be a lateral equivalent of facies D. This preferred succession suggests that although abrupt marine lithofacies contacts caused by storm events may be present, they are superimposed onto a fifth-order cyclic record. Non-marine facies were not statistically analysed because (i) their occurrence is limited to major unconformities separating fourth-order cycles, and (ii) fluctuating base-level seems unlikely to be clearly expressed in a succession of non-marine facies.

Based on both the physical features of each facies and their preferred progradational succession, depositional environments are interpreted to have gradationally ranged from open marine (facies A and B) through a shoal-bar complex (C, D and E) to lagoonal (E), to shoreface and foreshore (F and G) and thence to a mosaic of aeolian, fluvial and lacustrine environments (Table 1).

CYCLICITY

Thus far, our discussion of upper Hermosa cyclicity has been limited to a demonstration of the preferred Waltherian facies succession within fifth-order cycles. The presence of these cycles suggests potential for a

Table 2. Calculated binomial probabilities (applying the procedure of Harper, 1984) of observed gradational facies transitions for positive values within the probability difference matrix

Facies transition	N	n	p	q	Binomial probability
A–B	9	9	0.187	0.813	2.796×10^{-7}
B–C	21	3	0.083	0.917	0.251
B–D	21	8	0.278	0.722	0.206
C–E	1	1	0.732	0.268	0.732
D–E	17	7	0.375	0.625	5.735×10^{-8}
D–G	17	9	0.350	0.650	0.099
E–F	6	3	0.118	0.882	0.025
E–G	6	2	0.275	0.725	0.524
F–G	2	2	0.255	0.745	0.065
G–D	1	1	0.179	0.821	0.179

Table 3. (a) Transition count matrix, (b) observed transition probabilities, (c) transition probabilities for a random sequence and (d) probability difference matrices generated through Markov chain analysis; the high number of row/column mismatches is because only gradational facies transitions within fifth-order cycles are considered; analysis was limited to such successions, because marine facies contacts at abrupt fifth-order boundaries may represent non-contiguous depositional environments juxtaposed due to rapid base-level rise

(a) Transition count matrix

	A	B	C	D	E	F	G	Row sum	Row/column mismatch
A		9						9	9
B			3	8	7	2	1	21	12
C					1			1	2
D					7	1	9	17	7
E				1		3	2	6	9
F							2	2	13
G				1				1	4
Column sum	0	9	3	10	15	6	14	57	56

(b) Observed transition probabilities

	A	B	C	D	E	F	G
A		1					
B			0.14	0.38	0.33	0.10	0.05
C					1		
D					0.41	0.06	0.53
E				0.17		0.50	0.33
F							1
G				1			

(c) Transition probabilities for a random sequence

	A	B	C	D	E	F	G
A		0.19	0.06	0.21	0.31	0.12	0.29
B	0		0.08	0.28	0.42	0.17	0.39
C	0	0.16		0.18	0.27	0.11	0.25
D	0	0.22	0.07		0.37	0.15	0.35
E	0	0.18	0.06	0.20		0.12	0.27
F	0	0.16	0.05	0.18	0.27		0.25
G	0	0.16	0.05	0.18	0.27	0.11	

(d) Probability difference matrix

	A	B	C	D	E	F	G
A		+0.81	−0.06	−0.21	−0.31	−0.12	−0.29
B			+0.06	+0.10	−0.09	−0.07	−0.34
C		−0.16		−0.18	+0.73	−0.11	−0.25
D		−0.22	−0.07		+0.04	−0.09	+0.18
E		−0.18	−0.06	−0.03		+0.38	+0.06
F		−0.16	−0.05	−0.18	−0.27		+0.75
G		−0.16	−0.05	+0.82	−0.27	−0.11	

cyclic hierarchy in which fifth-order cycles stack into transgressive and regressive sets within fourth-order cycles (Vail *et al.*, 1977a; Posamentier *et al.*, 1988; Posamentier & Vail, 1988; Van Wagoner *et al.*, 1988). This stratigraphic framework was applied to the upper Hermosa to detect the hierarchy and mechanism(s) of cyclicity.

It is important to point out that the cyclic hierarchy

promoted by the previously cited sequence stratigraphers is a qualitative hierarchy, while placing an 'order' to the cycles is time-implicit (Miall, 1990). However, the amount of time represented by the upper member of the Hermosa is very poorly constrained. The ordering of upper Hermosa cycles presented in this study is, therefore, a qualitative (non-time-specific) ordering. The similarities between our hierarchical ordering of the upper Hermosa and the quantitative hierarchy described by Goldhammer *et al.* (1989) from the underlying Paradox Member, however, suggest that the two units may record cycles of similar duration.

Sequence stratigraphy

Application of concept

In outcrop and the subsurface, cyclicity within the upper Hermosa is pronounced, consisting of fourth-order marine–non-marine couplets ranging in thickness from 8 to 46 m (Fig. 5). In a general sense, Loope (1985) considered the marine beds a product of sea level rise, and the non-marine beds a product of low stand.

Recognition of deepening (retrogradational) and shallowing (progradational) fifth-order cycle sets requires assignment of relative water depths to each facies (Table 1). The direction of sea level change (i.e. rising or falling) between fourth-order cycles was determined semi-quantitatively: numbers were assigned to express the relative water depths of each facies in the previously described ideal Walther's succession:

Deep A → B → D → E → F → G → H–K Shallow
(B → C → E; D → G)
1 2 3 4 5 6 7

From this, transgressive (with each successive fifth-order cycle having a progressively deeper-water facies assemblage) and regressive (with each successive fifth-order cycle having a progressively shallower-water facies assemblage) fifth-order cycle

Fig. 5. Aeolian cross-strata resting directly on subtidal marine limestone (upper member, Hermosa Formation, approximately 24 km upstream from the mouth of Dark Canyon). Man is standing on knife-sharp flooding surface between oolitic grainstone (facies F) and shelf wackestone (facies B). At shoulder height, low-stand aeolian sandstone (facies H) abruptly overlies the transgressive fifth-order cycle set. Regressional fifth-order cycles succeeding facies B are not present, and are thought to have been removed by deflation.

sets were identified, and together define fourth-order cycles.

Fifth-order transgressive and regressive cycle sets within each fourth-order sequence were determined for all surface and subsurface sections (Figs 6 & 7). Uninterrupted successions of non-marine facies (aeolian, coastal plain (fluvial and/or loess) and lacustrine) abruptly overlying marine transgressive or regressive deposits were interpreted as low stand. As will be discussed later, this designation is tentative for cycles capped by coastal plain facies, inasmuch as Vail *et al.* (1977b) imply that inter-bedded coastal plain and marine deposits represent juxtaposed environments gradationally inter-bedded in response to sea level change. This suggests that coastal plain facies could be progradational, regressive sediments deposited before maximum low stand.

Differentiation of fifth-order transgressive and regressive cycle sets was difficult as well. This was partly because no more than five, and frequently as few as one, fifth-order cycles were observed within any given fourth-order cycle. In instances where only one fifth-order marine cycle was recognized, the cycle is sandwiched between non-marine facies (Figs 6 & 7), suggesting erosional truncation. The fewer the fifth-order cycles, the more difficult it is to recognize whether a cycle set is retrogradational or progradational. When only one or two fifth-order cycles were observed overlying low-stand deposits, they typically displayed only a retrogradational to aggradational stacking pattern. Such instances clearly imply transgressive deposition without either preservation or deposition of the succeeding regressive deposits (Figs 5 & 8a). Regressive deposits are difficult to recognize, because they commonly exhibit only an early high-stand aggradational stacking pattern without a succeeding progradational stacking of fifth-order cycle sets. There are intervals, however, where fifth-order stacking records a crudely symmetrical transgressive and regressive succession. In such instances, it was possible to differentiate the transgressive and regressive fifth-order cycle sets (Fig. 8b). When retrogradational fifth-order cycles were succeeded by only an aggradational fifth-order cycle set, the boundary between transgressive and regressive deposition was tentatively placed between the last retrogradational and first aggradational fifth-order cycle.

Mechanisms and hierarchy of cyclicity

A recurring theme of sequence stratigraphy is that cyclicity results from accommodation changes which may record eustatic sea level change (Kendall & Lerche, 1988). For the cause of cyclic changes in global sea level during the Late Pennsylvanian of the Mid-continent of North America, Wanless & Shepard (1936) and Heckel (1980) suggested the waxing and waning of Gondwana glaciers. The relative tectonic quiescence of the North American craton during Late Pennsylvanian deposition, and the high-frequency 'cyclothems' described by Heckel (1986) (twenty-five cycles during the Missourian and Virgilian), support this glacial hypothesis. The hypothesis is consistent with independent evidence on the magnitude and periodicity of Gondwana glaciation (Frakes, 1979; Crowell, 1982; Chumakov, 1985; Veevers & Powell, 1987).

The Paradox Basin during the Late Pennsylvanian is similar to the Mid-continent in that it, too, was the locus of marine deposition, and therefore should record similar high-frequency, glacio-eustatically driven cycles. Uplift of the Ancestral Rockies and subsidence of the Paradox Basin were greatest during Desmoinesian time, but continued through the end of the Pennsylvanian (Mallory, 1975; Stone, 1977; Frahme & Vaughn, 1983; Kluth, 1986). There were, thus, at least two sources of sedimentary accommodation.

Evidence within the upper Hermosa suggests that glacial eustasy, rather than jerky subsidence, induced fifth- and fourth-order cyclicity. Because the Paradox Basin was tectonically active during the Late Pennsylvanian, it might seem that instantaneous subsidence, as suggested by Cisne (1986), could have caused episodic base-level rise and, therefore, deposition of fifth-order cycles. This is consistent with the variable thicknesses of fifth-order cycles; they do not show a predictable increase in thickness upsection within transgressive deposits in response to a progressive increase in accommodation, or a similar decrease in thickness upsection within the regressive deposits in response to a decrease in accommodation. The subsidence hypothesis, however, fails to explain the repetitive, orderly succession of retrogradational and aggradational to progradational fifth-order cycles. Jerky subsidence would induce only a base-level rise. As a result, carbonate sediments would aggrade to the raised base-level, and would be gradationally succeeded by progradational tidal flat facies. Shelf facies would not become desiccated or exposed to oxidizing meteoric water. Exposure surfaces (fourth-order sequence boundaries) abruptly separate deeper-water

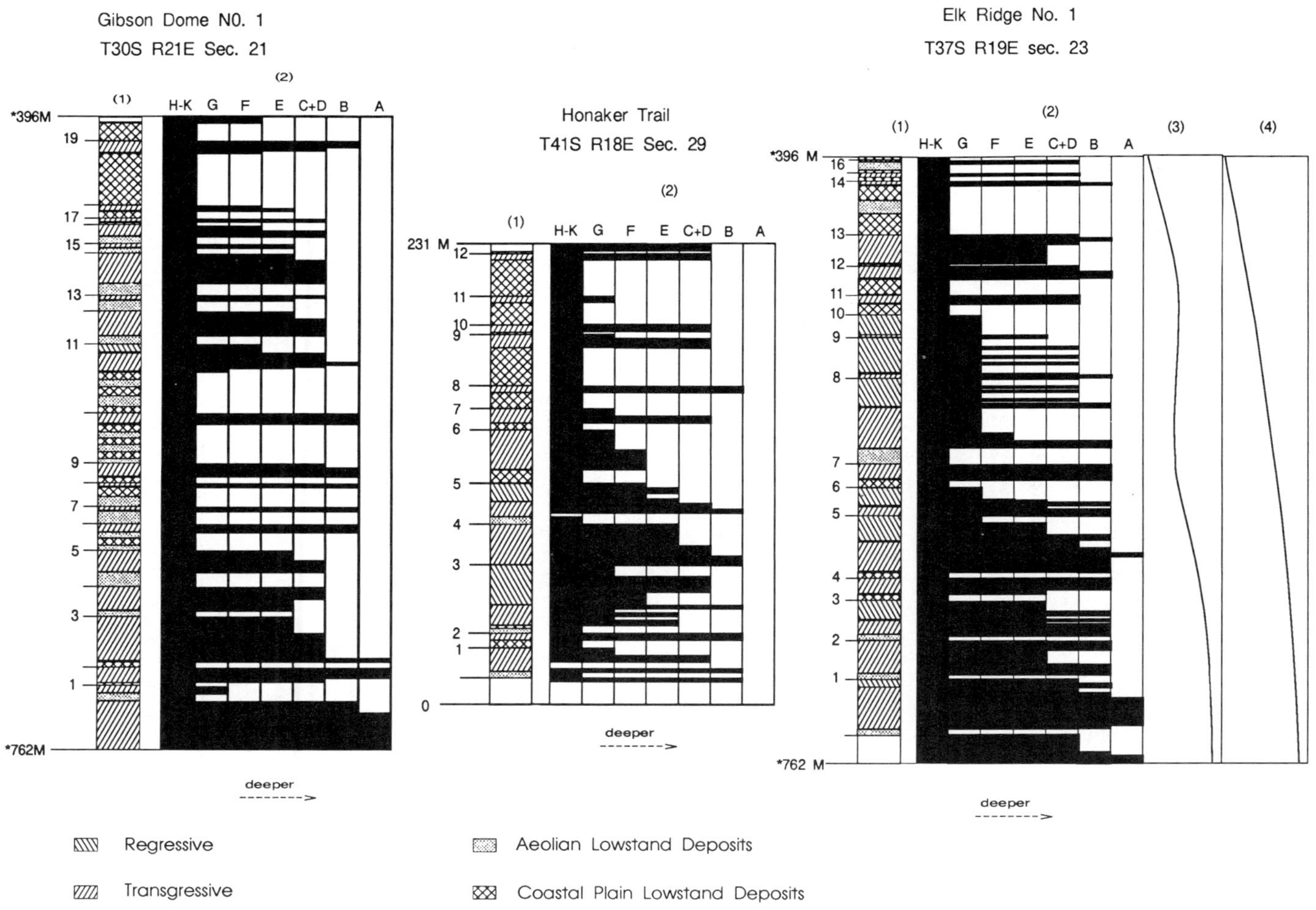

Fig. 6. Stratigraphic columns for Gibson Dome No. 1, Honaker Trail and Elk Ridge No. 1. Recorded in column (1) for each section is the vertical distribution of transgressive, regressive and low-stand deposits, along with the fourth-order cycles in which they occur (the numbers used to identify each fourth-order cycle are unique to each measured section). Note that the Honaker Trail section and Elk Ridge core have the most complete marine record (i.e. fourth-order cycles with both transgressive and regressive deposits preserved). These more 'complete' fourth-order cycles are almost always capped by terrestrial coastal plain to fluvial low-stand deposits. Column (2) is a graph of relative sea level generated by assigning relative water depth values to facies H/I/J/K (shallow) through to A (deep), and then shading in the corresponding value for each facies documented, such that the further to the right the shading occurs, the deeper the relative water depth that is recorded. From this semi-quantitative graph of relative sea level, it appears that the fifth- and fourth-order cycles shown in columns (2) and (1) are in turn superimposed on the third- and second-order cycles shown in columns (3) and (4) respectively.

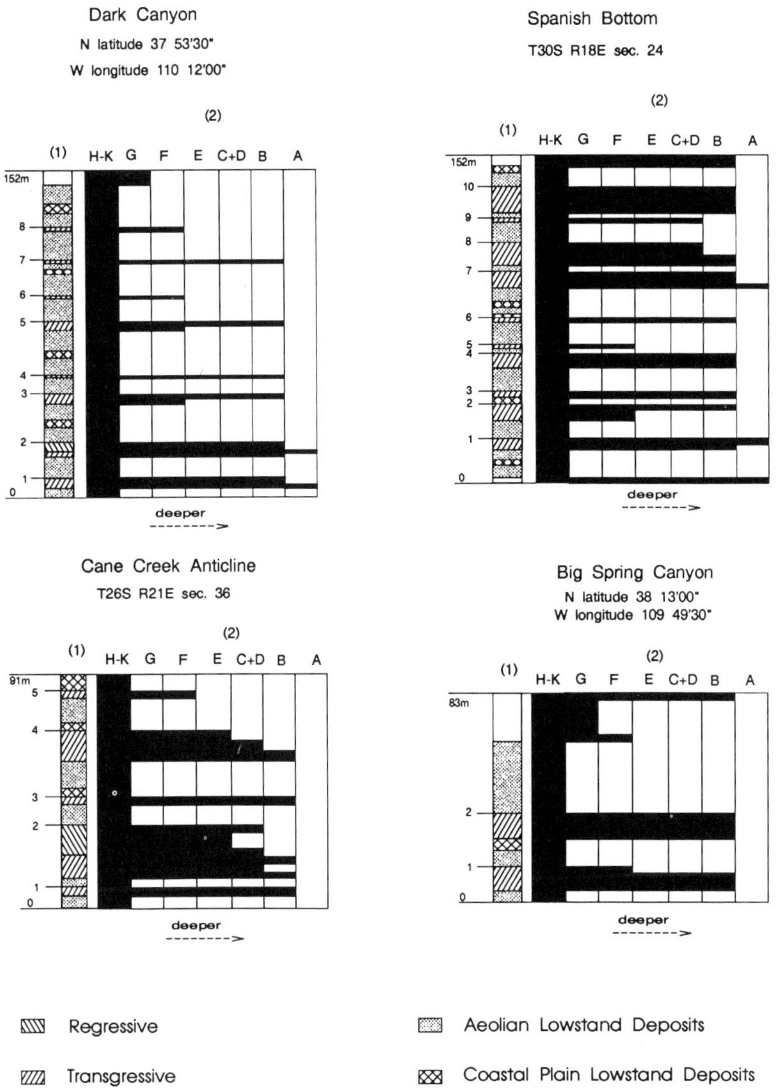

Fig. 7. Stratigraphic columns for Dark Canyon, Spanish Bottom, Cane Creek Anticline and Big Spring Canyon. Column (1) shows the distribution of transgressive, regressive and low-stand deposits along with the fourth-order cycles in which they occur (the numbers used to identify each fourth-order cycle are unique to each measured section). Unlike the sections at Honaker Trail and Elk Ridge (Fig. 6), marine units at these sections are very thin and rarely have transgressive marine deposits succeeded by marine regressive deposits. Transgressive deposits are nearly always abruptly overlain by thick aeolian low-stand deposits.

marine facies from overlying terrestrial low-stand deposits (Figs 5–7). Enos & Perkins (1977), Beach (1982) and Hardie (1986) indicate that exposure surfaces capping Quaternary carbonate shelf sequences result from high-amplitude glacio-eustatic sea level fluctuations. This is because the rate of sea level lowering exceeds the rate of tidal flat progradation (Hardie *et al.*, 1986; Read *et al.*, 1986). As a result, an exposure surface is formed with shelf marine units abruptly overlain by terrestrial low-stand deposits. The giant desiccation fissures described by Loope & Haverland (1988) and the evidence for oxidation of pyritized fossils (Loope & Watkins, 1989) from the upper Hermosa marine units are clear evidence not only of subaerial exposure, but also of sea level fall.

Because it lay beyond the trough axis from the Uncompahgre front, the Cane Creek Anticline section (Fig. 2) should have been insulated from the influx of clastics, yet be in a position where the effects of subsidence would be maximized. Low-amplitude sea level oscillations produced by jerky subsidence would have allowed the accumulation of relatively complete marine cycles. High-amplitude sea level oscillations, on the other hand, would have brought about total withdrawal of the sea

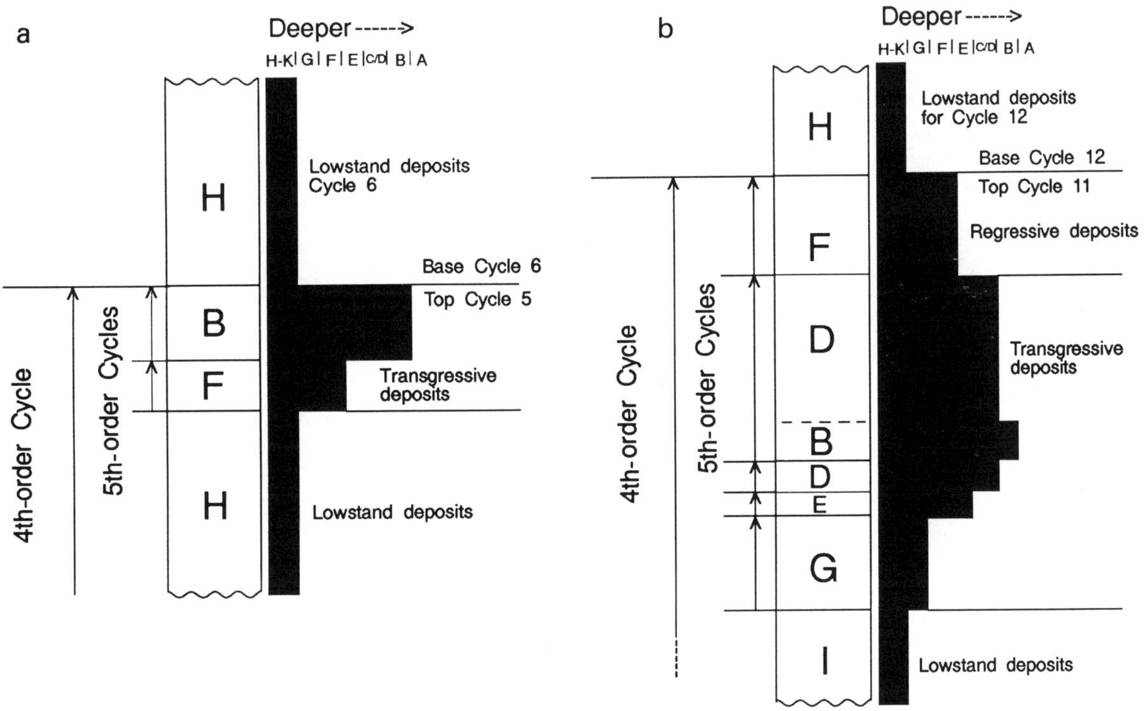

Fig. 8. (a) Example of an incomplete fourth-order cycle (cycle 5; Dark Canyon). In this instance, the aeolian dominated low-stand deposits are abruptly overlain by two transgressive fifth-order cycles, each of which is composed of a single facies. The lack of succeeding progradational marine fifth-order cycles suggests that regressive deposits were either never deposited, or were subsequently eroded prior to deposition of overlying low-stand aeolianites of fourth-order cycle 6 (see Fig. 7). (b) Example of a complete (and rarely observed) fourth-order cycle (cycle 11; Gibson Dome). (——) Knife-sharp contacts; (– – –) gradational boundary.

(even though tectonic accommodation was at a maximum), thereby exposing marine units to terrestrial erosional and depositional processes. The section is dominated by terrestrial low-stand deposits (Fig. 7, Table 4), attesting to the dominance of high-amplitude, glacio-eustatic sea level fluctuations.

From these observations, it appears that tectonic accommodation space was added so slowly that the influence on fifth- and fourth-order cycles was negligible. If so, then it seems that tectonic cyclicity should be manifested at a lower order. Goldhammer *et al.* (1989) suggest that sequences within the subjacent Paradox Member of the Hermosa occur in a shallowing and deepening set which defines a 200 m thick, third-order cycle. Based on observation of the graph of relative water depth for the Gibson Dome, Elk Ridge and Honaker Trail sections, we identified a 213 m third-order cycle which is similar in thickness to the third-order cycles of Goldhammer *et al.* (1989) in the underlying Paradox Member of the Hermosa Formation (Fig. 6).

It is possible that third-order cyclicity is a basinward manifestation of episodic tectonic pulses which Mack & Rasmussen (1984) suggest led to deposition of alluvial megacycles within the Permo-Pennsylvanian Cutler Formation on the eastern margin of the Paradox Basin. Third-order cyclicity may be attributed to either a tectonic or a long-term glacial mechanism (Miall, 1990). The only convincing evidence of a tectonic influence is the observation that the third-order deepening event at the top of the Gibson Dome and Elk Ridge cores consists of fifth- and fourth-order cycles dominated by shallower-water facies than the third-order deepening event at the base of these sections (Fig. 6). Furthermore, the third- and fourth-order cycles described by Goldhammer *et al.* (1989) in the underlying Paradox Member are dominated by marine facies, while the upper Hermosa becomes progressively dominated by terrestrial low-stand deposits upsection (Fig. 6). From this, it appears that third-order cycles within the Paradox Member and upper Hermosa may be

Table 4. Total number of metres (percentages in parentheses) occupied by each facies for each section; totals marked with asterisk do not include covered intervals or zones of missing core; sections include Spanish Bottom (SB), Elk Ridge (ER), Cane Creek Anticline (CC), Big Spring Canyon (BSC), Honaker Trail (HT), Dark Canyon (DC) and Gibson Dome (GD)

Facies	SB	ER	CC	BSC	HT	DC	GD
A	9.1 (6)	3.0 (<1)	—	—	—	3.1 (2)	29.2 (8)
B	50.2 (33)	102.5 (28)	14.6 (16)	19.9 (24)	15.1 (7)	13.7 (9)	36.6 (10)
C	—	22.0 (6)	—	—	—	—	—
D	7.6 (5)	22.0 (6)	16.4 (18)	—	39.0 (18)	1.5 (1)	76.9 (21)
E	—	47.6 (13)	10.0 (11)	—	17.3 (8)	—	32.9 (9)
F	6.1 (4)	1.5 (<1)	2.7 (3)	4.2 (5)	26.0 (12)	3.0 (2)	11.0 (3)
G	—	73.2 (20)	—	12.5 (15)	32.5 (15)	7.6 (5)	7.3 (2)
H	73.0 (48)	25.6 (7)	36.4 (40)	39.8 (48)	8.7 (4)	107.9 (71)	69.6 (19)
I	4.6 (3)	65.9 (18)	4.5 (5)	6.6 (8)	77.8 (36)	12.2 (8)	69.5 (19)
J	1.5 (<1)	2.7 (<1)	6.4 (7)	—	—	—	29.3 (8)
K	—	—	—	—	—	3.0 (2)	3.7 (1)
Total	152	366*	91	83	216.4*	152	366*

superimposed on a longer-term, tectonically induced second-order cycle (Fig. 6). This interpretation follows those of Mallory (1975), Frahme & Vaughn (1983) and Kluth (1986), who suggest maximum uplift and subsidence of the Ancestral Rockies during deposition of the Paradox Member (Middle Pennsylvanian time), followed by a progressive decrease in tectonism during deposition of the upper Hermosa and Rico (Late Pennsylvanian to Early Permian).

BEHEADED SEQUENCE HYPOTHESIS

The lateral continuity of depositional cycles or individual components of depositional cycles can be an important clue to the control of cyclicity. Because he could trace individual thin black shales for hundreds of kilometres, Heckel (1977) argued that the depositional cycles containing the shales must have been controlled by glacial eustasy, not by a more local mechanism like delta-switching. Based on similarities in the relative sea level curve generated for the upper Hermosa (Fig. 6), it is possible to correlate both the second-order episode of basin fill and the third-order cycles between the most continuous sections (Gibson Dome, Elk Ridge and Honaker Trail).

The fourth- and fifth-order cycles are anomalous for two reasons. Firstly, they are extremely hard to correlate, even on a local scale. Lewis & Campbell (1965) noted the lenticular nature of upper Hermosa limestones. Within our Honaker Trail and Elk Ridge sections, which are only 44 km apart, only one-third to one-quarter of the fourth-order sequences from Elk Ridge can be correlated (with marginal confidence) to sequences in the Honaker Trail section (Fig. 9). Correlation of fourth- and fifth-order cycles was accomplished by looking for similarities in the shape of the semi-quantitative graph of relative sea level along corresponding portions of the third-order sea level curve generated for the most continuous sections (Figs 6 & 9).

Secondly, fourth-order marine deposits are absent from many sequences (Figs 5, 6 & 7). Only the Elk Ridge and Honaker Trail sections have a significant number of fourth-order cycles with regressive marine deposits preserved. Fourth-order cycles at Gibson Dome, Big Spring Canyon, Cane Creek, Spanish Bottom and Dark Canyon have virtually no regressive deposits. Furthermore, not only do the Dark Canyon and Spanish Bottom sections generally lack regressive marine deposits, they also have thin, deeper-water transgressive deposits (Fig. 7). This is interesting considering that the Elk Ridge and Honaker Trail sections have coastal plain (facies I) dominated low-stand deposits, while the Big Spring Canyon, Cane Creek Anticline, Spanish Bottom and Dark Canyon sections have aeolian dominated low-stand deposits. Sections dominated by coastal plain low-stand deposits (fluvial transport processes during subaerial exposure) have the most complete marine record, while sections dominated by aeolian low-stand deposits (aeolian transport processes during subaerial exposure) have the most incomplete marine record (Figs 5, 6 & 7; Table 4). Explanations must be provided for the distribution of regressive marine and low-stand non-marine deposits, and for our inability to correlate many of the fourth-order cycles within the basin.

Shallow marine facies C–G, which were for the most part deposited above wave base, are less abundant in sections with aeolian dominated low-stand deposits (Table 4). Goldhammer *et al.* (1987) described a thick sequence of Triassic carbonates in which cycles were capped by subtidal facies that had been strongly altered by meteoric diagenesis. The

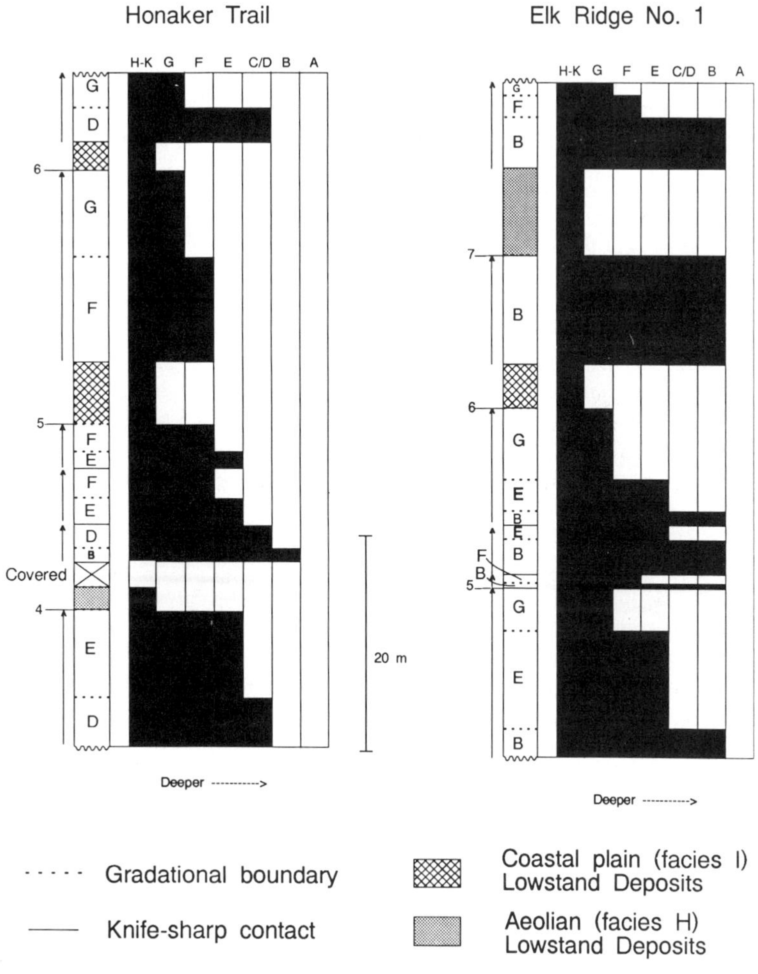

Fig. 9. Detailed stratigraphic columns for portions of Elk Ridge and Honaker Trail sections, showing correlations of fifth- and fourth-order cycles from a basin margin (Honaker Trail) to a more basinward (Elk Ridge) position. With regard to these sections, only fourth-order cycles 4, 5, 6 and 7 from Honaker Trail can be correlated confidently with cycles 5, 6, 7 and 8, respectively, from Elk Ridge. Note the deeper-water facies assemblage within the basinward Elk Ridge section.

lack of peritidal facies was attributed to the high amplitude of eustatic sea level change: shorelines migrated faster than coastal sediments could prograde. The lack of fourth-order regressive deposits in our basin-ward sections might be due to a similar mechanism. Unlike the Triassic example, however, palaeosols indicative of long episodes of subaerial exposure are not present at the tops of our fourth-order cycles. The lack of such palaeosols in our sections and the abrupt contacts between subtidal marine rocks and overlying aeolian strata suggest that erosion may have removed part of the marine record prior to deposition of the aeolian strata. Patterson & Kinsman (1981) stress the importance of the coastal hydrological regime in the preservation of Holocene tidal flat mudstones and evaporites marginal to the Persian Gulf. They indicate that progradational sediments deposited during the last 4000–5000 yrs which are above the regional water table are being stripped by wind erosion. We argue that the importance of deflation would be much greater during periods of eustatic sea level fall.

It seems likely that if high-amplitude sea level oscillations outpaced coastal plain (facies I and J) progradation, and if the continued fall of sea level allowed the groundwater table to drop below the sediment surface, marine carbonates would have been vulnerable to deflation. Sand-sized marine carbonates would have been stripped from the central and western portion of the basin and would have eventually accumulated downwind as low-stand aeolianites. Areas along the southern basin margin had marine high-stand deposits that were covered by an armouring mantle of coastal plain sediments soon after exposure (Fig. 9). Aside from explaining the lack of palaeosols, the scoured aspect of the cycle tops and the distribution of preserved marine and aeolian sands, this would also help to explain the high percentage of marine allochems in the aeolian deposits.

In an attempt to test this hypothesis, we analysed the vertical succession of facies. Miall & Gibling (1978) suggest that erosional contacts within a stratigraphic succession may be identified through the generation of a 'contact matrix' (Table 5). Values in the contact matrix range from +1 to −1, with increasingly negative scores indicating an increased facies association with scouring, and positive scores indicating non-erosional boundary conditions. Column sums represent the total erosional or gradational nature of a given facies to underlying facies, such that the more negative the column sum for a given facies, the more underlying facies it forms an erosional contact with. Aeolian facies H has the most negative column sum (Table 5) and therefore may have been deposited on an erosional to non-depositional surface. Facies I and B also have high negative values. The negative column sum for facies B is a response to disconformably juxtaposed facies along fifth-order flooding surfaces (Figs 5 & 8a & b), while the negative column sum for facies I is a result of deposition onto an erosional to non-depositional surface. As will be demonstrated, however, we consider fluvial erosion of marine strata to have been less significant than wind deflation, and, in fact, may have enhanced the preservation of underlying marine units.

Mud-rich, cohesive sediments can be deflated (Patterson & Kinsman, 1981), but probably not as readily as non-cohesive (sand-rich, granular) sediments. Facies A, B and E have a high percentage of interstitial mud (micrite), while facies C, D, F and G are mud-poor to mud-free (Table 1, Figs 3 & 4). Wind reworking implies that exposed marine shelf sediments were unlithified. Purser (1978) and Hird & Tucker (1988) suggest that a closed fabric, which is recognized by fitted, sutured and crushed grains, indicates that early cements were absent. Facies C, D, F and G were analysed to determine if they have an open or closed fabric. All shallow marine facies, except facies F (ooid grainstone), have a closed fabric and are thus interpreted to have been unconsolidated upon exposure. According to Ward (1973) and Wright (1988) the volume of early cements that develops in marine carbonates during episodes of subaerial exposure is climate dependent: the more water that moves through the pores, the more cement is emplaced. Previously reported observations of the upper Hermosa (Loope, 1985, 1988; Loope & Haverland, 1988; Loope & Watkins, 1989), as well as those made during this study, are consistent with a Late Pennsylvanian arid palaeoclimate for the Paradox Basin. The open fabric of facies F could be a result of marine cementation. The upper surface of one ooid grainstone within the upper Hermosa at Dark Canyon is clearly a hardground; it is encrusted by tabulate corals and is penetrated by abundant borings (Loope & Boyd, 1991).

In the upper Hermosa, the most complete marine record (where fourth-order cycles with both transgressive and regressive deposits are preserved) is found along the southern margin of the Paradox Basin at the Elk Ridge and Honaker Trail sections.

Table 5. Transition counts of (a) the gradational ($\mathbf{G}_{ij}$) and (b) the erosional ($\mathbf{E}_{ij}$) matrices used in construction of (c) the contact matrix ($\mathbf{C}_{ij}$); aeolian facies H has the most negative ($\mathbf{C}_{ij}$) column sum, indicating it is the facies which most frequently forms sharp (erosional) contacts with underlying units; based on these column sums, the following erosional hierarchy of facies, listed with increasing association with sharp lower boundaries, was generated: G, C, E, F, K, A, D, J, I, B, H; abrupt contacts associated with terrestrial facies H, I and J are probably a manifestation of both wind and water erosion, while abrupt contacts associated with marine facies are probably a reflection of abruptly juxtaposed facies at fifth-order cycle boundaries

(a) Gradational transition matrix ($\mathbf{G}_{ij}$)

	A	B	C	D	E	F	G	H	I	J	K
A	0	9	0	0	0	0	0	0	0	0	0
B	0	0	3	8	7	2	1	0	0	0	0
C	0	0	0	0	1	0	0	0	0	0	0
D	0	0	0	0	7	1	9	0	0	0	0
E	0	0	0	1	0	3	2	2	0	0	0
F	0	0	0	0	0	0	2	0	0	0	0
G	0	0	0	1	0	0	0	0	1	0	0
H	0	0	0	0	0	0	0	0	2	0	0
I	0	2	0	1	0	0	0	5	0	0	0
J	0	0	0	0	0	0	0	0	2	0	0
K	0	0	0	0	0	0	0	0	0	0	0
Total = 72											

(b) Erosion transition matrix ($\mathbf{E}_{ij}$)

	A	B	C	D	E	F	G	H	I	J	K
A	0	1	0	0	0	0	0	2	0	0	0
B	5	0	0	2	1	1	0	10	10	2	0
C	0	2	0	0	0	0	0	2	0	0	0
D	1	4	0	0	1	1	2	4	8	0	0
E	0	5	0	3	0	0	2	3	1	1	0
F	0	3	0	3	1	0	0	4	1	1	0
G	0	3	0	5	3	0	0	0	6	0	0
H	2	15	0	5	2	5	0	0	15	5	2
I	1	9	0	9	4	2	4	12	0	5	0
J	0	1	0	1	0	1	0	4	3	0	1
K	0	0	0	0	0	0	0	2	1	0	0
Total = 215											

(c) Contact matrix ($\mathbf{C}_{ij}$)

	A	B	C	D	E	F	G	H	I	J	K
A		+0.8						−1			
B	−1		+1	+0.6	+0.7	+0.3	+1	−1	−1	−1	
C		−1			+1			−1			
D	−1	−1			+0.7		+0.6	−1	−1		
E		−1		−0.5		+1		−0.2	−1	−1	
F		−1		−1	−1		+1	−1	−1	−1	
G		−1		−0.7	−1				−0.7		
H	−1	−1		−1	−1	−1			−0.8	−1	−1
I	−1	−0.6		−0.8	−1	−1	−1	−0.4		−1	
J		−1		−1		−1		−1	−0.2		−1
K								−1	−1		
Column sum	−4	−6.8	+1	−4.4	−1.6	−1.7	+1.6	−7.6	−6.7	−5.0	−2

Here, regressive deposits succeeded by low-stand exposure are invariably capped with fluvial facies I. Erosion at times preceded facies I deposition as can be seen at the Honaker Trail, Gibson Dome and Elk Ridge sections where transgressive deposits are abruptly overlain by fluvial facies I low-stand deposits without any evidence of an intervening regressive episode (Fig. 6). Furthermore, facies I and J most often form disconformable contacts with underlying marine units (Table 5), and occasionally have carbonate bioclasts and lithoclasts as a detrital sedimentary component (Fig. 10a & b).

North of the Elk Ridge and Honaker Trail sections, the lack of aggradational coastal plain deposits apparently left shelf carbonate sediments exposed to subaerial processes. In these sections, there is a lack of both coastal plain and shallow marine deposits, suggesting that low-stand processes preceding coastal plain deposition along the southern margin of the Paradox Basin were not as erosive as the low-stand processes preceding aeolian deposition within the basin centre (Table 4, Fig. 7). A troubling point is determining how it is possible to deposit and preserve thick successions of fine-grained fluvial to aeolian (loessite) sediments during low stand of both sea level and the terrestrial water table. It seems these sediments would be reworked by the wind during episodes of base-level (and water table) lowering. An alternative explanation would be to consider the coastal plain facies at Honaker

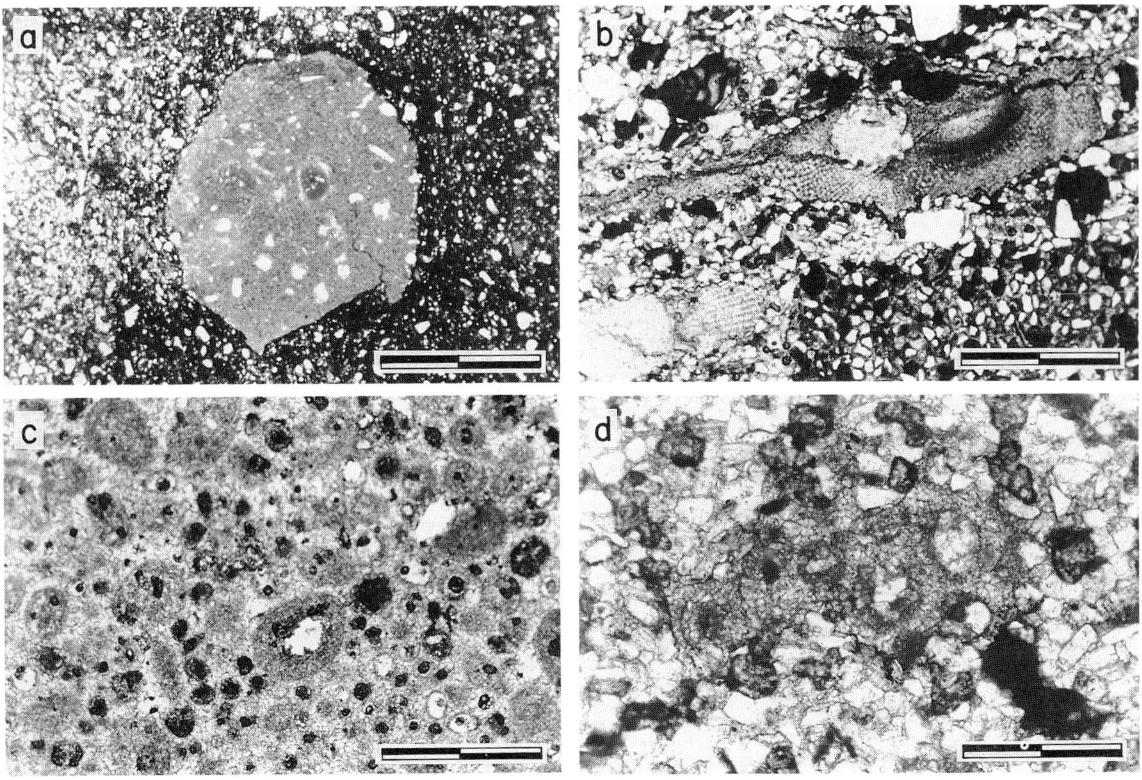

Fig. 10. Photomicrographs. Bar scale = 1 mm. (a) Facies I (coastal plain); Elk Ridge core at 415 m below ground level. The large carbonate lithoclast in this sample was incorporated into facies I through erosion of marine facies E. (b) Facies J (coastal plain); Gibson Dome core at 459 m below ground level. Note abundant carbonate bioclasts and lithoclasts that were eroded from marine strata. (c) Facies F (marine); Gibson Dome core at 527 m below ground level. This grainstone facies comprises a relatively thin succession of regressive marine deposits (cycle 11). Note that this ooid grainstone is abruptly overlain by a low-stand aeolianite (Fig. 6). (d) Facies H (aeolian); Gibson Dome core at 527 m below ground level. The lithoclast in the centre of the photograph looks identical to the underlying facies F grainstone shown in (c), suggesting it was reworked from this regressive unit. Note that even though the underlying grainstone has an open fabric (suggesting early induration), erosional processes were still able to rework the unit.

Trail and Elk Ridge as part of the aggradational to progradational sedimentary wedge deposited during middle to late high stand. This interpretation makes it easier to explain accretion of the fine-grained coastal plain sediments through the presence of a sediment binding, high water table, and further allows aggradational veneering of underlying marine high-stand deposits, thereby protecting them from low-stand erosion. Although this explanation is plausible, it is contradicted by several lines of evidence. Firstly, if the coastal plain sediments along the southern margin of the basin were deposited during sea level high stand, then they should be succeeded stratigraphically by deposits (or surfaces) towards the basin centre which are low-stand. Only the Gibson Dome site appears transitional between coastal plain and aeolian deposition (Fig. 6). Here, regressive marine deposits do not give way upsection to coastal plain and then aeolian deposits. Rather, coastal plain and aeolian deposits are randomly inter-bedded, suggesting autocyclic deposition during terrestrial exposure. With regard to this observation, it is interesting to consider the work of Johnson (1989), which suggests that Middle Pennsylvanian to Early Permian Maroon Formation loessites along the southern margin of the Eagle Basin (Central Colorado Trough) were derived from contemporaneous dunes found to the north within the basin centre. It seems the successions of facies I along the southern margin of the Paradox Basin could be a low-stand analogue to Johnson's observations. Secondly, if coastal plain facies were part of a regressive succession, then individual fifth-order cycles would gradationally become more dominated by coastal plain facies as these cycles stacked towards the succeeding fourth-order cycle boundary. In the sections studied, however, regressive fifth-order cycles consist solely of marine facies abruptly terminated by coastal plain facies rather than becoming progressively dominated by coastal plain facies (Fig. 6). Furthermore, transgressive marine deposits were at times abruptly overlain by coastal plain facies without an intervening shoaling marine facies succession (Fig. 6). Finally, marine bioclasts and lithoclasts are a fairly common detrital component in coastal plain sediments, thereby suggesting base-level lowering and erosion of an exposed marine shelf (Fig. 10).

The beheaded sequence hypothesis suggests that our inability to correlate fourth-order cycles may be the result of partial or total deflation of marine carbonate units during fourth-order low stands of sea level. Implicit in this hypothesis is that the components of 'missing' carbonate facies should occur as clasts within low-stand aeolianites. To test this idea, marine facies were grouped into those of relatively deep-water (facies A and B), or shallow-water (facies C–G) origin. Facies percentages for these categories were calculated in each section and compared to the percentage of aeolian and fluvial facies (Fig. 11). Sections dominated by low-stand aeolian deposits generally lack shallow-water facies. To determine if shallow marine sediments occur as detritus in low-stand aeolianites, ninety-one aeolianite samples were point-counted to determine the mean percentage of recognizable carbonate bioclasts and lithoclasts. The observed percentages ranged from 33% to less than 1%, having a mean value of 11%. These are minimum values, because only recognizable carbonate grains were counted (many grains were squashed beyond recognition). Carbonate grains within aeolianites are dominated by bioclasts from facies D and G, but also include lithoclasts attributed to facies B, E and F (Figs 4b & 11). This suggests that although B, E and F were lithified, the subaerial processes acting on these units could nonetheless promote disintegration and deflation.

Based on the observations presented, it appears that two end-members of stratigraphic completeness are manifested in the upper member of the Hermosa Formation of the Paradox Basin (Fig. 12). Along the southern margin of the basin in the vicinity of the Elk Ridge and Honaker Trail sections, the transgressive and regressive marine carbonate fifth-order components of fourth-order cycles were frequently preserved. This preservation may have been enhanced by an armouring mantle of alluvium or loess during fourth-order low stands. Within the more central part of the basin to the north, deflation stripped a portion or all of the transgressive and regressive marine carbonate record. Subsequently, correlation of fourth-order cycles across the basin is difficult.

SUMMARY AND CONCLUSIONS

1 Detailed facies analysis of seven surface and subsurface sections of Upper Pennsylvanian rocks (upper member of the Hermosa Formation) within the Paradox Basin reveals that eleven marine and non-marine depositional environments are represented.

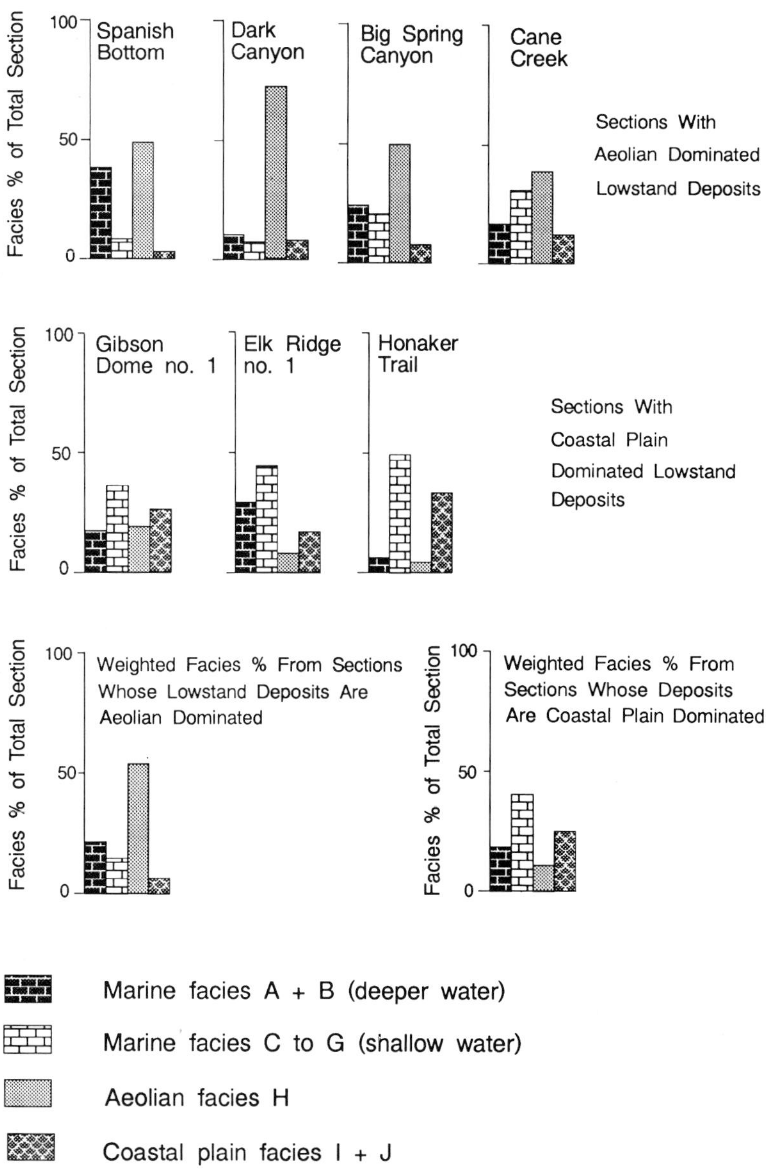

Fig. 11. Facies percentages for each section studied. The total percentage of section occupied by relatively deep-water marine facies, and relatively shallow-water facies, were plotted with the percentage of section occupied by aeolian facies, and coastal plain facies. Sections were separated into those with aeolian dominated low-stand deposits, and those with coastal plain dominated low-stand deposits. For these two categories, the weighted percentages of deep and shallow marine, and coastal plain and aeolian, facies were determined. Observation of the facies percentages indicates that, with an increase in aeolian dominance, there is an associated decrease in shallow marine facies. Conversely, with an increase in coastal plain dominance, there is an increase in the percentage of shallow marine deposits.

2 Markov chain analysis of gradational facies contacts within fifth-order cycles allows reconstruction of the progradational succession of marine facies. Through consideration of both the physical features of the facies and their ideal Waltherian succession within a fifth-order cycle, it was determined that the marine facies were deposited on a barred shelf landward from which was a terrestrial mosaic of aeolian, lacustrine and fluvial environments.

3 Using the relative water depth assemblage for each fifth-order cycle, a hierarchy of stratigraphic cyclicity can be recognized within the upper member of the Hermosa Formation. Within this framework, fifth-order cycles occur in deepening (transgressive) and shallowing (regressive) sets to form fourth-order cycles, which in turn are grouped into a third-order deepening and shallowing event. Finally, this third-order cycle is interpreted to be superimposed on a second-order episode of basin fill. High-amplitude glacio-eustatic sea level oscillations are probably

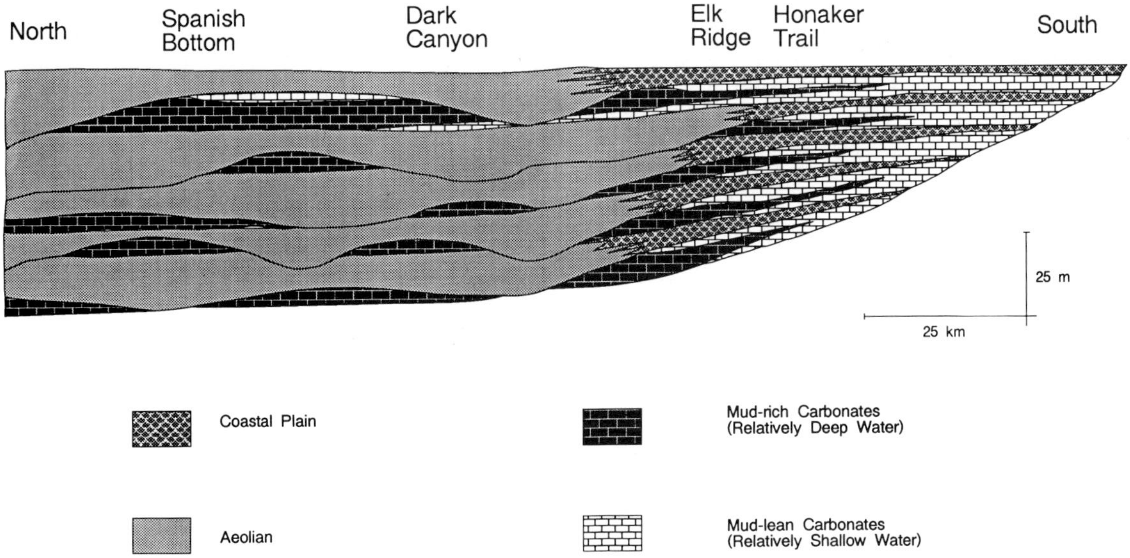

Fig. 12. Diagrammatic cross-section showing configuration of facies and deflation surfaces for five hypothetical fourth-order cycles. To the north, much of the marine portion of fourth-order cycles was reworked by the wind into low-stand aeolianites. Correlation of northern truncated cycles with the relatively more complete cycles to the south would be difficult. Note extreme vertical exaggeration (×1000).

responsible for creating the fifth- and fourth-order cycles. Variations in subsidence rate generated the second-, and perhaps the third-order cycles.

4 Although fourth-order cyclicity is attributed to an allocyclic mechanism (glacio-eustasy), few individual cycles can be correlated across the basin. Fourth-order cycles appear to be truncated in that transgressive deposits are rarely succeeded by aggradational to progradational deposits. Those cycles with an abbreviated marine history are usually overlain by aeolian low-stand deposits, which contain significant amounts of 'missing' shallow-water carbonate grains. High-amplitude sea level oscillations left these shallow marine deposits subaerially exposed, where they could be deflated and reworked into aeolian low-stand deposits. Correlation of these allocyclic events is therefore difficult.

ACKNOWLEDGEMENTS

Funded by NSF Grant EAR-8816222. Materials and data were provided by Exxon Company, USA, South-western Division, and D.H. Buckner of Denver, Colorado. R.F. Diffendal, J.B. Swinehart and W.J. Wayne (University of Nebraska), R.K. Goldhammer and S.R. Schutter (Exxon Production Research) and K. Havholm and R.P. Langford (University of Texas) reviewed the manuscript and offered very constructive criticism.

REFERENCES

Beach, D.K. (1982) *Depositional and diagenetic history of Pliocene–Pleistocene carbonates of northwestern Great Bahama Bank; evolution of a carbonate platform.* Unpublished PhD thesis, University of Miami.

Blakey, R.C. (1980) Pennsylvanian and early Permian paleogeography, southern Colorado Plateau and vicinity. In: *Paleozoic Paleogeography of the West-central United States* (Ed. Fouch, T.D. & Magathan, E.R.) pp. 239–257. Soc. Econ. Paleont. Mineral., Rocky Mountain Section, Rocky Mountain Paleogeography Symposium 1.

Boardman II, D.R. & Heckel, P.H. (1989) Glacial–eustatic sea level curve for early Late Pennsylvanian sequence in north-central Texas and biostratigraphic correlation with curve for midcontinent North America. *Geology* **17**, 802–805.

Boardman II, D.R. & Malinky, J.M. (1985) Glacial–eustatic control of Virgilian cyclothems in north central Texas. *Am. Ass. Petrol. Geol., Southwest Section Meeting Trans.*, 13–23.

Busch, R.M.Q. & Rollins, H.B. (1984) Correlation of Carboniferous strata using a hierarchy of transgressive–regressive units. *Geology* **12**, 471–474.

Caputo, M.V. & Crowell, J.C. (1985) Migration of glacial centers across Gondwana during Paleozoic Era. *Bull. Geol. Soc. Am.* **96**, 1020–1036.

Chumakov, N.M. (1985) Glacial events of the past and their geological significance. *Palaeogeog., Palaeoclimatol., Palaeoecol.* **51**, 319–346.

Cisne, J.L. (1986) Earthquakes recorded stratigraphically on carbonate platforms. *Nature* **323**, 320–322.

Crowell, J.C. (1982) Continental glaciation through geologic times. *Climate in Earth History* pp. 77–82. National Academy Press.

Dalrymple, R.W., Narbonne, G.M. & Smith, L. (1986) Eolian action and the distribution of Cambrian shales in North America. *Geology* **13**, 607–610.

Driese, S.G. & Dott, R.H., Jr (1984) Model for sandstone–carbonate 'Cyclothems' based on upper member of Morgan Formation (Middle Pennsylvanian) of northern Utah and Colorado. *Bull. Am. Ass. Petrol. Geol.* **68**, 574–597.

Enos, P. (1983) Shelf. In: *Carbonate Depositional Environments* (Ed. Scholle, P.A., Bebout, D.G. & Moore, C.H.) pp. 267–295. Mem. Am. Ass. Petrol. Geol. 33.

Enos, P. & Perkins, R.D. (1977) Quaternary sedimentation in south Florida. *Mem. Geol. Soc. Am. 147*, 198 pp.

Fairbridge, R.W. & Johnson, D.L. (1978) Eolianite. In: *The Encyclopedia of Sedimentology* (Ed. Fairbridge, R.W. & Bourgeois, J.) pp. 279–283. Dowden, Hutchinson & Ross, Stroudsburg.

Frahme, C.W. & Vaughn, E.B. (1983) Paleozoic geology and seismic stratigraphy of the northern Uncompahgre front, Grand County, Utah. In: *Rocky Mountain Foreland Basins and Uplifts* (Ed. Lowell, J.D.) pp. 201–211. Rocky Mountain Association of Geologists.

Frakes, L.A. (1979) *Climates throughout Geologic Time*. Elsevier, New York, 310 pp.

Goldhammer, R.K., Dunn, P.A. & Hardie, L.A. (1987) High-frequency glacio-eustatic sea level oscillations with Milankovitch characteristics recorded in Middle Triassic platform carbonates in northern Italy. *Am. J. Sci.* **287**, 853–892.

Goldhammer, R.K., Oswald, E.J. & Dunn, P.A. (1989) The hierarchy of stratigraphic forcing—an example from middle Pennsylvanian (Desmoinsean) shelf carbonates of the southwest Paradox Basin, Honaker Trail, Utah [abstract]. In: *Sedimentary Modeling: Computer Simulation of Depositional Sequences* (Ed. Franseen, E.K. & Watney, W.L.) pp. 27–30. Subsurface Geology Series 12, Kansas Geological Survey.

Goldstein, R.H. (1988) Cement stratigraphy of Pennsylvanian Holder Formation, Sacramento Mountains, New Mexico. *Bull. Am. Ass. Petrol. Geol.* **72**, 425–438.

Goodwin, P.W. & Anderson, E.J. (1985) Punctuated aggradational cycles: a general hypothesis of cyclic stratigraphic accumulations. *J. Geol.* **93**, 515–533.

Hardie, L.A. (1986) Stratigraphic models for carbonate tidal-flat deposition. *Colorado School of Mines Quarterly* **81**, 59–74.

Hardie, L.A., Bosellini, A. & Goldhammer, R.K. (1986) Repeated subaerial exposure of subtidal carbonate platforms, Triassic, northern Italy: evidence for high frequency sea level oscillations on a 10^4 year scale. *Paleoceanog.* **1**, 447–457.

Harper, C.W., Jr (1984) Improved methods of facies sequence analysis. In: *Facies Models*, 2nd edition (Ed. Walker, R.G.) pp. 11–13. Reprint Series 1, Geoscience Canada.

Heckel, P.H. (1977) Origin of phosphatic black shale facies in Pennsylvanian cyclothems of Midcontinent North America. *Bull. Am. Ass. Petrol. Geol.* **61**, 1045–1068.

Heckel, P.H. (1980) Paleogeography of eustatic model for deposition of midcontinent upper Pennsylvanian Cyclothems. In: *Paleozoic Paleogeography of the West-central United States* (Ed. Fouch, T.D. & Magathan, E.R.) pp. 197–215. Soc. Econ. Paleont. Mineral., Rocky Mountain Section, Rocky Mountain Paleogeography Symposium 1.

Heckel, P.H. (1986) Sea-level curve for Pennsylvanian eustatic marine transgressive–regressive depositional cycles along midcontinent outcrop belt, North America. *Geology* **14**, 330–334.

Hird, K. & Tucker, M.E. (1988) Contrasting diagenesis of two Carboniferous oolites from south Wales: a tale of climatic influence. *Sedimentology* **35**, 587–602.

Hite, R.J. (1960) Stratigraphy of the saline facies of the Paradox Member of the Hermosa Formation of southeastern Utah and southwestern Colorado, in geology of the Paradox Basin fold and fault belt. *Four Corners Geological Society Guidebook, 3rd Field Conference*, 86–89.

Hite, R.J. & Buckner, D.H. (1981) Stratigraphic correlations, facies concepts, and cyclicity in Pennsylvanian rocks of the Paradox Basin. *Rocky Mountain Association of Geologists, 1981 Field Conference*, 147–159.

Johnson, D.L. (1968) Quaternary eolianites: do they have ancient analogs [abstract]? *Spec. Pap. Geol. Soc. Am.* **121**, 150–151.

Johnson, S.Y. (1989) Significance of loessite in the Maroon Formation (Middle Pennsylvanian to Lower Permian), Eagle Basin, northwest Colorado. *J. Sed. Petrol.* **59**, 782–791.

Kendall, C.G. St C. & Lerche, I. (1988) The rise and fall of eustasy. In: *Sea-level Changes: an Integrated Approach* (Ed. Wilgus, C.K. *et al.*) pp. 3–18. Spec. Publ. Soc. Econ. Paleont. Mineral. 42, Tulsa.

Kluth, C.F. (1986) Plate tectonics of the Ancestral Rocky Mountains. In: *Paleotectonics and Sedimentation* (Ed. Peterson, J.A.) pp. 353–369. Mem. Am. Ass. Petrol. Geol. 41.

Koerschner III, W.F. & Read, J.F. (1989) Field and modeling studies of Cambrian carbonate cycles, Virginia Appalachians. *J. Sed. Petrol.* **59**, 654–687.

Kozar, M.G., Weber, L.J. & Walker, K.R. (1990) Field and modelling studies of Cambrian carbonate cycles, Virginia Appalachians—discussion. *J. Sed. Petrol.* **60**, 790–794.

Lewis, R.Q. & Campbell, R.H. (1965) Geology and uranium deposits of Elk Ridge and vicinity, San Juan County, Utah. *US Geol. Surv. Prof. Pap. 474-B*, 69 pp.

Loope, D.B. (1984) Eolian origin of upper Paleozoic sandstones, southeastern Utah. *J. Sed. Petrol.* **54**, 563–580.

Loope, D.B. (1985) Episodic deposition and preservation of eolian sands: a late Paleozoic example from southeast Utah. *Geology* **13**, 73–76.

Loope, D.B. (1988) Rhizoliths in ancient eolianites. In: *Late Paleozoic and Mesozoic Eolian Deposits of the Western Interior of the United States* (Ed. Kocurek, G.) pp. 301–314. Sed. Geol. 56.

Loope, D.B. & Boyd, D.W. Cemented sea-floors of Pennsylvanian age, southeastern Utah. *Geol. Soc. Am.*,

Abstr. Programs **23**, A66.

LOOPE, D.B. & HAVERLAND, Z.E. (1988) Giant desiccation fissures filled with calcareous eolian sand, Hermosa Formation (Pennsylvanian), southeastern Utah. In: *Late Paleozoic and Mesozoic Eolian Deposits of the Western Interior of the United States* (Ed. Kocurek, G.) pp. 403–413. Sed. Geol. 56.

LOOPE, D.B. & WATKINS, D.K. (1989) Pennsylvanian fossils replaced by red chert: early oxidation of pyritic precursors. *J. Sed. Petrol.* **59**, 375–386.

MACK, G.H. & RASMUSSEN, K.A. (1984) Alluvial fan sedimentation of the Cutler Formation (Permo-Pennsylvanian) near Gateway, Colorado. *Bull. Geol. Soc. Am.* **95**, 109–116.

MALLORY, W.M. (1975) Middle Southern Rocky Mountains, northern Colorado Rocky Mountain Plateau and eastern Great Basin. *US Geol. Surv. Prof. Pap. 853*, 263–278.

MIALL, A.D. (1990) *Principles of Sedimentary Basin Analysis*, 2nd edition. Springer-Verlag, New York, 668 pp.

MIALL, A.D. & GIBLING, M.R. (1978) The Siluro-Devonian clastic wedge of Somerset Island, Arctic Canada, and some regional paleogeographic implications. *Sed. Geol.* **21**, 279–286.

MIDDLETON, G.V. (1973) Johannes Walther's law of correlation of facies. *Bull. Geol. Soc. Am.* **84**, 979–988.

PATTERSON, R.J. & KINSMAN, D.J.J. (1981) Hydrologic framework of a sabkha along Arabian Gulf. *Bull. Am. Ass. Petrol. Geol.* **65**, 1457–1475.

POSAMENTIER, H.W., JERVEY, M.T. & VAIL, P.R. (1988) Eustatic controls on clastic deposition I—conceptual framework. In: *Sea-level Changes: an Integrated Approach* (Ed. Wilgus, C.K. *et al.*) pp. 109–124. Spec. Publ. Soc. Econ. Paleont. Mineral. 42.

POSAMENTIER, H.W. & VAIL, P.R. (1988) Eustatic controls on clastic deposition II—sequence and systems tract models. In: *Sea-level Changes: an Integrated Approach* (Ed. Wilgus, C.K. *et al.*) pp. 125–154. Spec. Publ. Soc. Econ. Paleont. Mineral. 42.

PURSER, B.H. (1978) Early diagenesis and the preservation porosity in Jurassic limestones. *J. Petrol. Geol.* **1**, 83–94.

READ, J.F., GROTZINGER, J.P., BOVA, J.A. & KOERSCHNER, W.F. (1986) Model for generation of carbonate cycles. *Geology* **14**, 107–110.

RICE, J.A. & LOOPE, D.B. (1991) Wind-reworked marine carbonates, late Paleozoic of northwest Arizona and southern Nevada. *Bull. Geol. Soc. Am.* **103**, 254–267.

SARG, J.F. (1988) Carbonate sequence stratigraphy. In: *Sea-level Changes: an Integrated Approach* (Ed. Wilgus, C.K. *et al.*) pp. 155–181. Spec. Publ. Soc. Econ. Paleont. Mineral. 42.

STONE, D.S. (1977) Tectonic history of the Uncompahgre Uplift. In: *Exploration Frontiers of the Central and Southern Rockies* (Ed. Veal, H.K.) pp. 23–30. Rocky Mountain Association of Geologists Guidebook.

VAIL, P.R., MITCHUM, R.M., JR & THOMPSON III, S. (1977b) Seismic stratigraphy and global changes of sea level, part 3: relative changes of sea level from coastal onlap. In: *Seismic Stratigraphy—Applications to Hydrocarbon Exploration* (Ed. Clayton, C.E.) pp. 63–81. Mem. Am. Ass. Petrol. Geol. 26.

VAIL, P.R., MITCHUM, R.M., JR, TODD, R.G., WIDMIER, J., THOMPSON III, S., SANGREE, J.B., BUBB, J.N. & HATELID, W.G. (1977a) Seismic stratigraphy and global changes of sea level, part 2: the depositional sequence as a basic unit for stratigraphic analysis. In: *Seismic Stratigraphy—Applications to Hydrocarbon Exploration* (Ed. Clayton, C.E.) pp. 53–62. Mem. Am. Ass. Petrol. Geol. 26.

VAN WAGONER, J.C., POSAMENTIER, H.W., MITCHUM, R.M., VAIL, P.R., SARG, J.F., LOUTIT, T.S. & HARDENBOL, J. (1988) An overview of the fundamentals of sequence stratigraphy and key definitions. In: *Sea-level Changes: an Integrated Approach* (Ed. Wilgus, C.K. *et al.*) pp. 39–45. Spec. Publ. Soc. Econ. Paleont. Mineral. 42.

VEEVERS, J.J. & POWELL, C.M. (1987) Late Paleozoic glacial episodes in Gondwanaland reflected in transgressive–regressive sequences in Euramerica. *Bull. Geol. Soc. Am.* **98**, 475–487.

WALKER, R.G. (1979) Facies and facies models. General introduction. In: *Facies Models*, 1st edition (Ed. Walker, R.G.) pp. 1–7. Reprint Series 1, Geoscience Canada.

WANLESS, H.R. & SHEPARD, F.P. (1936) Sea level and climatic changes related to late Paleozoic cycles. *Bull. Geol. Soc. Am.* **47**, 1177–1206.

WARD, W.C. (1973) Influence of climate on the early diagenesis of carbonate eolianites. *Geology* **1**, 171–174.

WENGERD, S.A. (1973) Pennsylvanian stratigraphy in the Paradox Basin, Four Corners region. In: *Pennsylvanian Systems in the US* (Ed. Branson, C.C.) pp. 264–330. Am. Ass. Petrol. Geol.

WRIGHT, V.P. (1988) Paleokarsts and paleosols as indicators of paleoclimate and porosity evolution: a case study from the Carboniferous of South Wales. In: *Paleokarst* (Ed. James, N.P. & Choquette, P.W.) pp. 329–341. Springer-Verlag, New York.

Spec. Publs Int. Ass. Sediment. (1993) **16**, 151–161

Draa reconstruction, the Permian Yellow Sands, northeast England

T. CHRINTZ* *and* L.B. CLEMMENSEN

Institute of General Geology, University of Copenhagen, Øster Voldgade 10, DK-1350 Copenhagen K, Denmark

ABSTRACT

The lower Permian Yellow Sands of northeast England have been preserved as a number of 1.5–3.5 km wide and *c.*30 m high, northeast–southwest trending linear draa ridges. The three-dimensional sedimentary build-up or aeolian architecture of these draa ridges is beautifully exposed in a number of large quarries. Studies of these outcrops allow a reconstruction of draa morphodynamics. The sediments in the draa ridges are composed of two superimposed inter-draa–draa element sets. The draa elements, which are the focus of this paper, are composed of abundant dune elements and minor inter-dune elements. The cross-stratified dune elements are between 0.5 and 11 m thick, with most elements having a thickness of between 2 and 4 m. In sections transverse to the draa ridge the dune elements are 2–40 m wide and have a trough-shaped lower bounding surface. The elements occur in both left-hand and right-hand versions and commonly form inter-digitating zones. In longitudinal sections the element has a planar lower bounding surface and appears to be of tabular geometry. The cross-beds of the dune elements display a squeezed bimodal pattern with one mode towards the northwest (313°) and the other mode towards the south-southeast (171°).

Computer-graphics modelling suggests that the linear draas were superimposed by sinuous linear dunes. These superimposed dunes were reversing and non-migrating (perfectly longitudinal) and had sinuosities that migrated along-crest. The draas and superimposed dunes were in equilibrium and apparently formed in a bimodal wind regime with alternating northwest or north winter winds and southeast or east summer winds. The resultant sand transport direction was parallel to the draa ridges, suggesting that these also were longitudinal bedforms during the final phase of draa sedimentation.

INTRODUCTION

The reconstruction of aeolian bedform type and behaviour from outcrop data has long interested sedimentologists, and the correct interpretation of ancient aeolian bedforms has important implications not only for palaeogeographic and palaeoclimatic reconstructions, but also for practical characterization of aeolian reservoirs.

The classic approach is to make comparative analyses of the sedimentary structures of a specific ancient deposit and of modern aeolian bedforms (e.g. McKee, 1979a, b; Tsoar, 1982). Additional methods involve studies of foreset bedding attitudes (e.g. Glennie, 1970; Ahlbrandt & Fryberger, 1980), and studies of the amount and distribution of aeolian stratification types (e.g. Hunter, 1981). The interpretation of ancient aeolian bedforms has been treated recently in much detail by Rubin (1987), who used computer programs to recreate the morphology and behaviour of the bedforms that produced a specific deposit. This latter approach is employed also in this study.

Such computer-aided reconstructions of aeolian morphodynamics are dependent on detailed field studies of the three-dimensional structural build-up, or aeolian architecture, of a specific deposit. Reconstruction of aeolian morphodynamics, therefore, is best done on high-quality three-dimensional exposures.

The Permian Yellow Sands of northeast England

* Present address: NSR Consult, Sortemosevej 2, DK-3450, Allerød, Denmark.

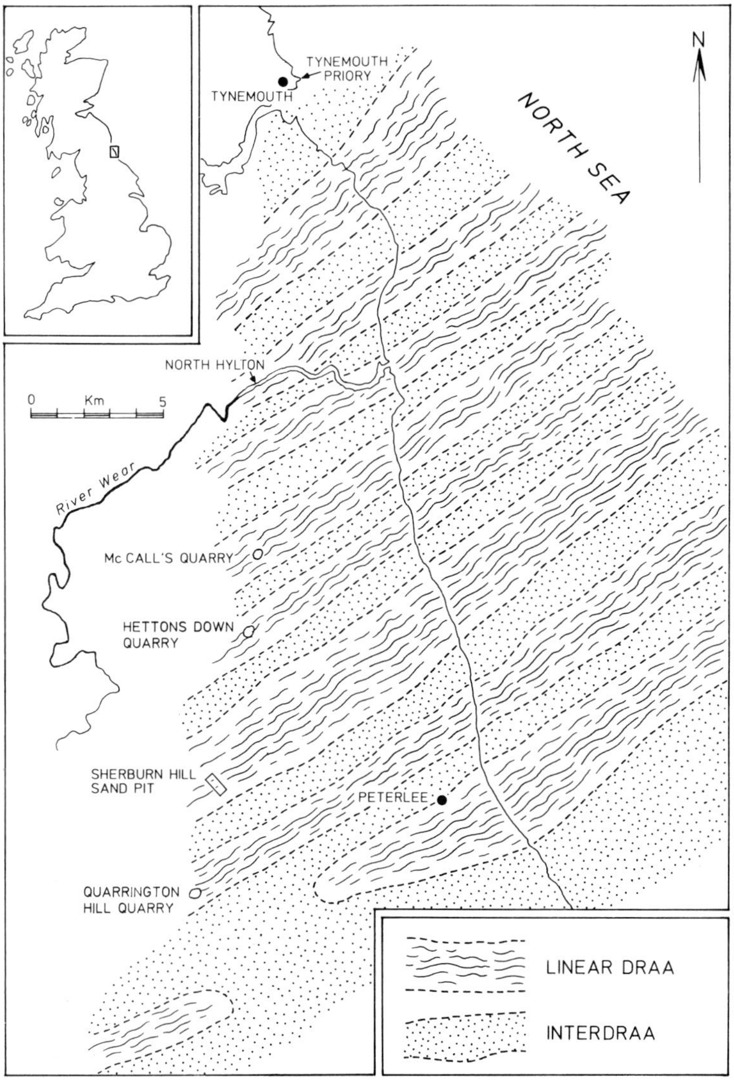

Fig. 1. Map showing the location of the quarries studied in the Yellow Sands. (After Clemmensen, 1989.)

are well exposed in three dimensions in a number of large, actively worked quarries near Durham (Fig. 1). Repeated visits to these quarries since 1984 have given us a good knowledge of facies and facies architecture in the Yellow Sands (e.g. Chrintz & Clemmensen, 1989, 1990; Clemmensen, 1989). In this study, we intend to describe the aeolian architecture of the Yellow Sands in some detail, with emphasis on the dominant trough-shaped (dune) element, and to reconstruct aeolian morphodynamics with the help of computer-graphics modelling. Our reconstructions indicate that the Yellow Sands represent large linear draas with superimposed sinuous linear dunes. The Yellow Sands thereby seem to constitute one of the very few examples of linear aeolian bedforms preserved in the ancient record (e.g. Rubin & Hunter, 1985).

GEOLOGICAL SETTING

The Yellow Sands are an aeolian formation of Permian age which outcrops extensively in northeast England (Fig. 1). The formation rests unconformably on Carboniferous strata and is overlain by Late Permian marine rocks (e.g. Smith & Francis, 1967). The Yellow Sands have been ascribed to the Weissliegend, which corresponds to the uppermost non-reddened part of the Rotliegend sandstone sequence (Glennie & Buller, 1983). The aeolian deposits are believed to represent part of a large erg at the northwestern margin of the Southern Permian Basin (e.g. Glennie, 1983).

The Yellow Sands have been preserved as several sand ridges up to 57 m high and 1.5–3.5 km wide, separated by 0.8–2 km wide corridors with a thin sand cover (Steele, 1983; Clemmensen, 1989). These sand ridges are thought to represent partially preserved aeolian morphology.

The present investigation is based primarily on studies of four large quarries in Durham: Quarrington Hill Quarry, Sherburn Hill Sand Pit, Hettons Down Quarry and McCall's Quarry (Fig. 1).

All of the quarries are situated within the sand ridges, but the relative position of the quarries varies from the ridge centre to the ridge margin.

The origin of the Yellow Sands has been the subject of much interest. Although an aeolian origin has been accepted by all but Pryor (1971), there has been no general agreement on the exact morphodynamics of the aeolian bedforms that produced the Yellow Sands (cf. Smith & Francis, 1967; Glennie, 1983; Glennie & Buller, 1983; Yardley, 1984; Mader & Yardley, 1985). The most recent view, expressed by Clemmensen (1989) and Chrintz & Clemmensen (1990), considers the bedforms to be large compound linear dunes or draas. In this paper we shall further document and extend this interpretation of the Yellow Sands.

AEOLIAN ARCHITECTURE

The three-dimensional sedimentary structure or aeolian architecture of the Yellow Sands is well exposed in the quarries studied (Fig. 2). A preliminary account of the aeolian architecture was given by Chrintz & Clemmensen (1989). A more comprehensive architectural element analysis is presented here based on the nomenclature in Clemmensen (1991a, b).

In general, aeolian architecture can be described on four levels: micro-scale, meso-scale, macro-scale and mega-scale sedimentary architecture (Table 1).

Meso-scale aeolian architecture refers to the amount and distribution of the basic aeolian stratification types, and will be briefly commented upon during the description of macro-scale aeolian architecture. The latter refers to the amount and distribution of the basic aeolian architectural elements,

Table 1. Aeolian architecture (after Clemmensen, 1990, 1991)

Scale	Basic components	Application
Micro-scale	Texture elements	Sorting processes
Meso-scale	Stratification types	Aeolian processes
Macro-scale	Architectural elements	Bedform dynamics
Mega-scale	Sedimentary storeys	Desert evolution

Table 2. Aeolian architectural elements (after Clemmensen, 1990, 1991)

Rank		
High-order	Intermediate-order	Low-order
Erg element		
	Draa element	
	Inter-draa element	
	Dune element	
	Inter-dune element	
		Stratification element
	Sand sheet element	
	Sabkha element	

which are: erg element, draa element, inter-draa element, dune element, inter-dune element, sand sheet element, sabkha element and stratification or bedform modification element (Table 2). The Yellow Sands contain all but the sabkha element (Fig. 3). Aeolian mega-scale architecture refers to the number and geometry of well defined erg units or aeolian storeys within the studied formation. In the Yellow Sands only one main erg unit seems to be presented, and the mega-scale architecture is therefore relatively simple. In the following only the macro-scale aeolian architecture will be described.

Macro-scale aeolian architecture

Erg element

Because the Yellow Sands are considered to represent only one erg unit, the erg element corresponds to all sediments in the formation. As already mentioned, this erg element constitutes several large sand ridges representing partly preserved linear draas.

It is evident from field studies that the erg element in each sand ridge is composed of two superimposed inter-draa–draa element sets overlain by a thin sand sheet element (Figs 2, 3 & 4; Clemmensen, 1989).

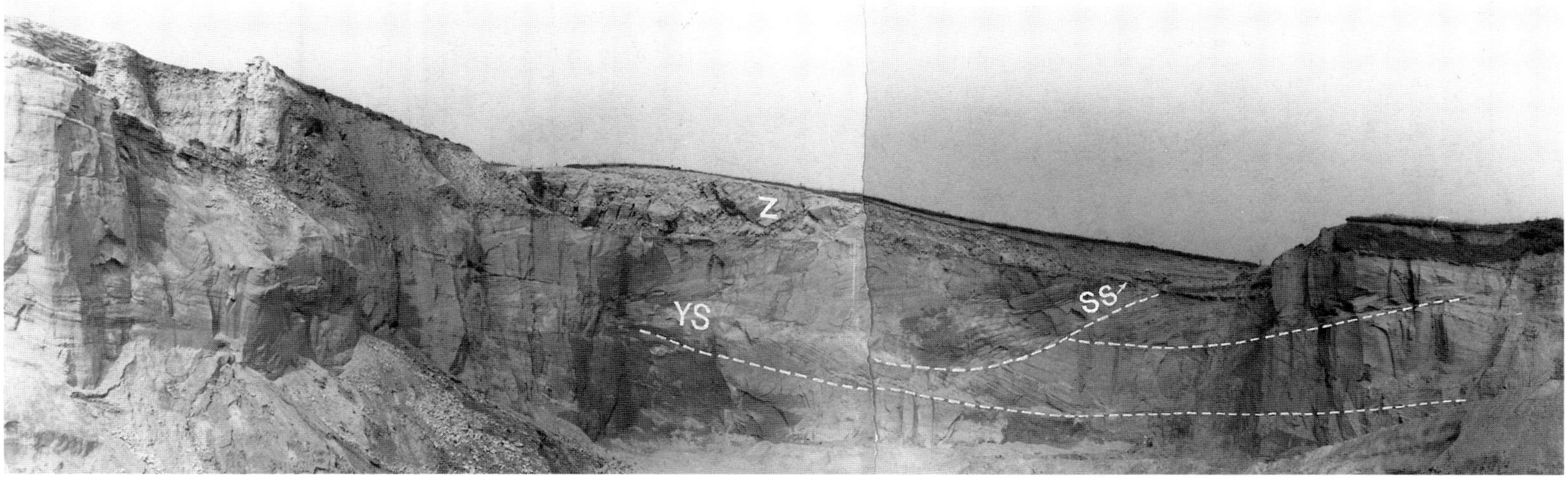

Fig. 2. Three-dimensional exposure of the Yellow Sands (YS) in McCall's Quarry. The far wall is orientated northwest–southeast, or nearly perpendicular to the sand ridge, while the nearest walls are both orientated nearly parallel to the sand ridge. The Yellow Sands are here represented by the upper draa element (exposed thickness *c*.25 m) overlain by a thin and sporadically preserved sand sheet element (SS). The draa element is composed of relatively large dune elements (up to 10 m thick) that are trough-shaped in the transverse section and of tabular geometry in the longitudinal sections. Discontinuous lines indicate lower bounding surfaces in selected elements. The Yellow Sands are overlain by marine Zechstein deposits (Z).

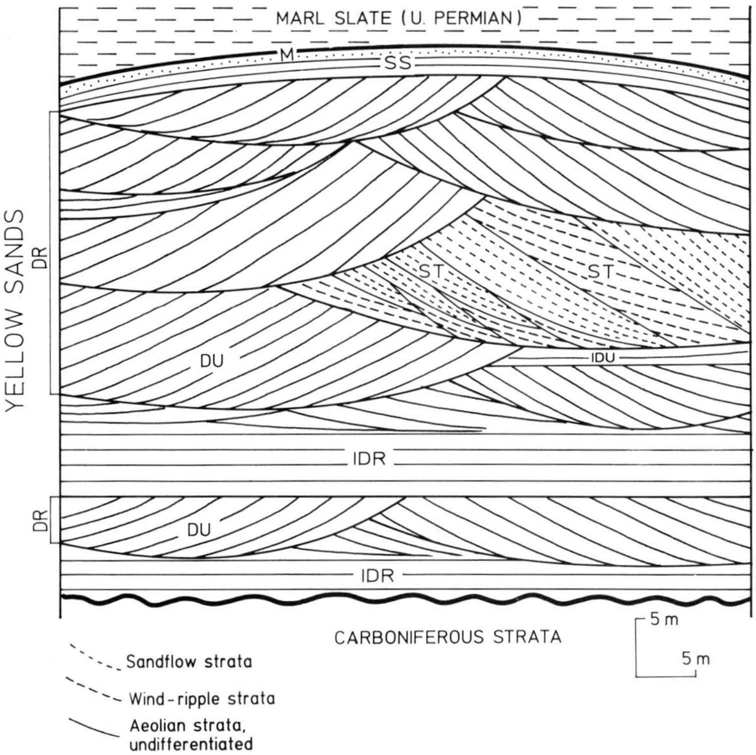

Fig. 3. Idealized cross-section of a sand ridge in the Yellow Sands illustrating the sedimentary architecture of the aeolian deposits. DR, Draa element; IDR, inter-draa element; DU, dune element; IDU, inter-dune element; ST, stratification element; SS, sand sheet element; M, sand reworked during Zechstein transgression.

Fig. 4. Part of erg element in the Sherburn Hill Sand Pit. The wall trends nearly perpendicular to the sand ridge and is composed of a lowermost flat-bedded to low-angle stratified inter-draa–draa plinth element (IDR), a middle and relatively thick draa element (DR) and an upper thin sand sheet element (SS). The upper convex-up surface of the Yellow Sands is thought to represent the original, somewhat subdued morphology of the draas.

The uppermost part of the sand sheet element shows evidence of marine reworking (Fig. 3).

The erosional surface between the uppermost draa element and the sand sheet element does not contain any features such as soil or bioturbation indicative of a large break in sedimentation. The sand sheet element therefore seems to indicate draa starvation possibly coupled to the early phase of marine transgression and a resulting decrease in sand supply.

Draa elements

Each sand ridge in the Yellow Sands is composed of two inter-draa–draa element sets separated by an inter-draa-order or first-order bounding surface (cf. Clemmensen, 1989). Both draa elements are underlain by an inter-draa element, but the preserved thickness and geometry of the two draa elements differ greatly. The lower draa element has a sheet-like geometry and a thickness of *c*.3 m. It formed during an episode of slow lateral migration of the draas. The upper draa element forms convex ridges up to 3000 m wide and 10–40 m high (Figs 3 & 4). This upper draa element formed primarily during an episode of vertical draa accretion, but draa sediments were eroded to some extent by wind prior to sand sheet formation. Both draa elements are composed mainly of trough-shaped dune elements. The description of draa architecture and morphodynamics refers mainly to the upper draa element (Figs 2, 4, 5 & 6).

Inter-draa elements

The sand ridges contain two inter-draa–draa plinth elements, but only the upper one is well exposed. This upper inter-draa element rests on an inter-draa-order surface, is of a sheet-like geometry and has a typical thickness of 5–9 m. The sedimentary features and genesis of this element have already been described in detail by Clemmensen (1989).

Dune elements

This element makes up *c.*90% of the upper draa element and is the most conspicuous element in the Yellow Sands; a correct interpretation of it is crucial to the understanding of the aeolian morphodynamics. This element is therefore described in more detail here.

The element is 2–40 m wide and 0.5–11 m thick and possesses an asymmetric trough-shaped lower bounding surface in transverse sections (Figs 2, 3 & 6). It contains slightly oblique cross-bedding, with sandflow strata in the more steeply dipping 'proximal' part, and wind-ripple strata in the less steeply or counter-dipping 'distal' part. The element appears in both right-hand and left-hand versions, which often results in the formation of inter-digitation zones. In longitudinal sections the element has a

Fig. 6. Dune elements exposed in transverse section, Hettons Down Quarry.

Fig. 5. Dune elements exposed in a longitudinal to oblique section, Hettons Down Quarry. Discontinuous lines define one element.

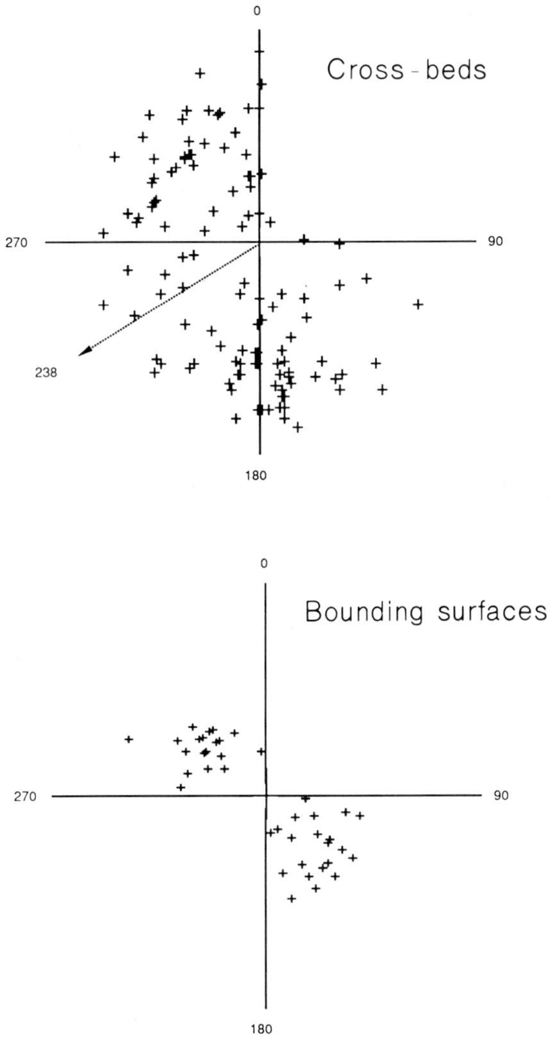

Fig. 7. Data on cross-bed and bounding surface orientation, dune elements, Yellow Sands. Data were collected in all the quarries examined.

rather planar lower bounding surface, and in these sections the element actually appears to be of tabular geometry (Fig. 5).

Measurements of foreset bedding attitudes yield a well ordered pattern, which can be classified as a squeezed bimodal compass rose (Fig. 7). The foresets dip up to 25–30°, with one mode towards the northwest (mean = 313°) and one mode towards the south-southeast (mean = 171°). The two modes are of nearly equal importance. Readings of the associated dune-order bounding surfaces also yield a bimodal dip distribution (Fig. 7). It should be noted, however, that the majority of the bounding surfaces only dip 2–10° and display a nearly perfect bipolar pattern with modes towards the northwest and southeast (see also Yardley, 1984).

The thickness of the element is relatively constant at any one locality (quarry), but element thickness varies systematically between the localities. Quarries situated near the edge of a sand ridge possess the thinnest elements (mean = 2 m), whereas quarries situated near a ridge centre have the thickest elements (mean = 5 m; maximum = 11 m).

The elements often display scalloped cross-bedding (cf. Rubin, 1987) as seen in both transverse and longitudinal sections. Each scallop, which is here termed a stratification element or bedform modification element, is bounded above and below by dipping bounding surfaces (Fig. 3).

The size of the trough-shaped element suggests that it represents aeolian bedforms (dunes) superimposed on the large linear draas. The geometry and internal distribution of aeolian strata suggest that the element records deposition on basal slip faces and in adjacent potholes or scour pits. The bimodal orientation of cross-beds and bounding surfaces indicates deposition on the opposed flanks of reversing, linear and longitudinal bedforms, and the occurrence of an along-crest component of dip to the cross-beds is explained by along-crest migration of sinuosities on the linear dune. This general interpretation of the element will be tested and further refined by computer-graphics modelling.

Inter-dune elements

This element makes up *c*.10% of the draa element, although it is scarcely present in some of the quarries situated near the ridge centre. The inter-dune–dune plinth element is of weakly wedge-shaped appearance in transverse sections. It is up to 3 m thick and up to 20 m wide, and composed of low-angle to horizontal wind-ripple strata. The element most commonly occurs as an isolated feature interbedded between thicker dune elements (Fig. 3). The element is thought to represent deposition in inter-dune corridors and on low-angle dune plinths associated with the superimposed linear dunes. Its rare presence is ascribed primarily to common wind erosion (scour formation) in the inter-dune areas.

BEDFORM RECONSTRUCTION

The aeolian bedforms that produced the Yellow Sands have previously been interpreted as simple linear dunes (seifs) (Smith & Francis, 1967; Glennie, 1970; Glennie & Buller, 1983), linear (and longitudinal) draas (Steele, 1983) and transverse draas (Yardley, 1984).

The large size of the sand ridges, their elongated form and consistent northeast–southwest orientation indicate that the Yellow Sands do represent linear bedforms of draa size (cf. Steele, 1983; Clemmensen, 1989). The ancient linear bedforms had an average width of 2150 m and an average spacing of 3100 m, and are thereby somewhat larger than the largest modern linear draas (see Lancaster, 1983, 1988). These ancient linear draas seem to have been oblique bedforms during an early phase of sedimentation, but true longitudinal bedforms during the final phase of draa sedimentation (Clemmensen, 1989).

The sedimentary build-up of the draa elements indicates the existence of superimposed aeolian bedforms. Steele (1983) interpreted these superimposed aeolian bedforms as crescentic transverse dunes, but computer simulation of bedforms comparable to those envisaged by Steele yields sedimentary structures and a cross-bed dip pattern different from those characterizing the Yellow Sands (see Rubin, 1987, fig. 56).

We interpret the superimposed aeolian bedforms as sinuous linear dunes, and from the architecture of the dune element we infer that these linear dunes were reversing (reversals in bedform asymmetry), longitudinal bedforms with sinuosities that migrated along-crest. The reconstruction of the aeolian bedforms is based on trial-and-error simulation of the structures observed in the field (cf. Rubin, 1987).

The results of one of these computer experiments are illustrated in Fig. 8. The sinuous longitudinal dunes in the model are characterized by vertical climbing (cf. Rubin, 1987, fig. 77). At the same time the draas were slowly migrating sidewards and eventually accreting vertically (cf. Clemmensen, 1989). The structures formed by the longitudinal dunes are trough-shaped in transverse sections, but of nearly tabular geometry in longitudinal sections. In transverse sections structures (or elements) form regularly spaced zones of inter-digitation in which

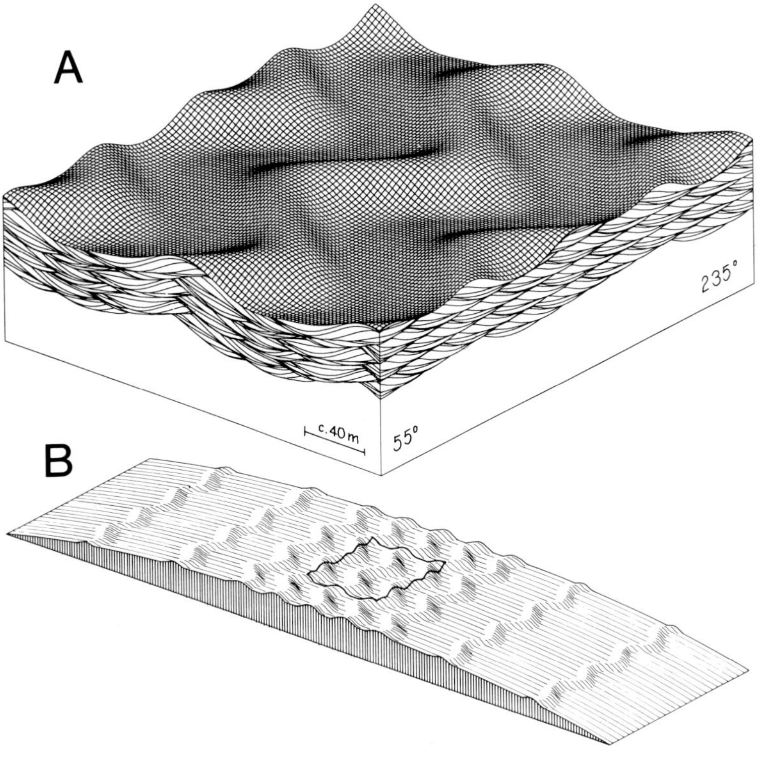

Fig. 8. Computer-graphics reconstruction of sedimentary structures produced by longitudinal reversing sinuous dunes (cf. Rubin, 1987, fig. 77). In the Yellow Sands these longitudinal dunes (A) were superimposed on the central part of the draa ridges (B). The structures produced can be viewed as a reasonable approximation to the real structures of the dune elements in the Yellow Sands. An even better match could probably result if the sinuous dunes were less regular, had peaks and saddles and were associated with shallow scour pits at the base of the dune slopes.

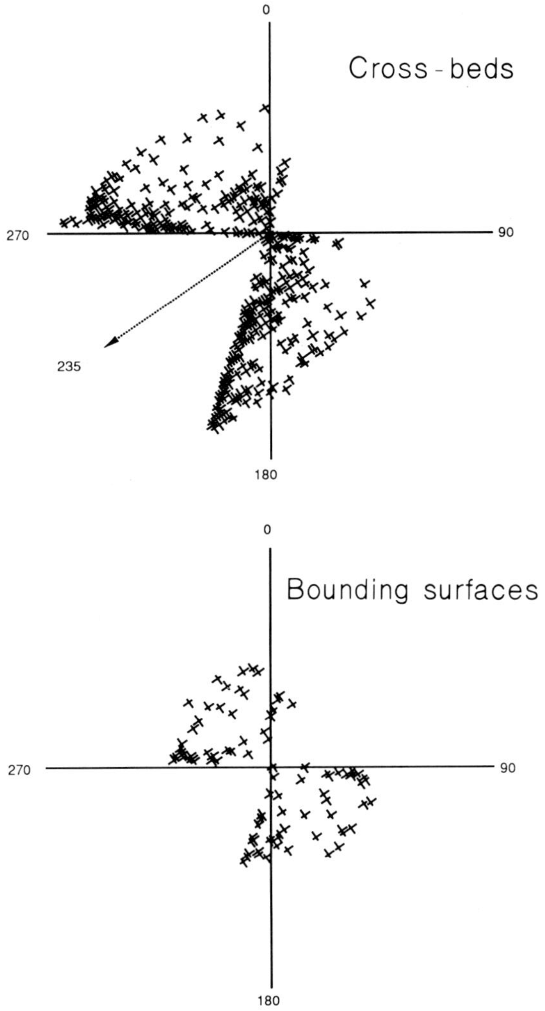

Fig. 9. Computer image of cross-beds and bounding surfaces produced by the bedforms in the theoretical model (Fig. 8).

left-hand and right-hand versions of the elements are almost symmetrically distributed on either side of a mirror plane. The elements are also characterized by scalloped cross-bedding. Finally the cross-bed plot of the bedforms examined is of a squeezed bimodal appearance (Fig. 9), with a marked downwind asymmetry. Thus in nearly all aspects the observed structures and the structures produced by the sinuous longitudinal dunes in the computer modelling are identical.

There are also some important differences, however, between the observed and the modelled structures. One is the relatively large amount of along-crest dipping cross-beds in the Yellow Sands. Evidently the cross-beds represent the topographically lowest portion of the superimposed dune–inter-dune system, and to judge from the geometry of these structures, the ancient inter-dune areas may frequently have contained shallow scour pits formed by secondary airflow.

The occurrence of wind scour is to be expected in inter-dune areas if they are not protected from deflation (Kocurek *et al.*, 1992). In the preferred reconstruction the inter-dune areas occurred between linear dunes superimposed on the main draa. It is likely therefore that the inter-dune areas were well above groundwater level, facilitating wind scour.

Another difference is the slight angular deviation between many cross-beds in the Yellow Sands and the related bounding surfaces, which is most apparent for data in the southeastern quadrant. This deviation may once again suggest the importance of wind scouring and successive filling of the depressions formed in combination with the overall along-crest migration of dune sinuosities.

Thus, based on outcrop data and computer-graphics modelling, it is concluded that the Yellow Sands represent large linear (oblique to longitudinal) draas with superimposed linear (longitudinal) dunes. Similar bedforms have rarely been recorded from other geological formations (Bell, 1991), but they are common in several modern sand seas, including the Namib Sand Sea (e.g. Lancaster, 1983), the Sahara (Breed *et al.*, 1979) and the Rub'al Khali of Saudi Arabia (Breed *et al.*, 1979). It is considered likely therefore that continued studies of aeolian formations will result in new finds of fossil compound linear or longitudinal dunes.

WIND REGIME

The large sand ridges (draas) trend almost exactly in the same direction (55–235°) as the resultant sand drift direction (238°). It is concluded therefore that the large linear bedforms during the final phase of draa sedimentation were longitudinal draas that were in equilibrium with the wind regime.

The genesis of longitudinal aeolian bedforms has been much discussed (e.g. Lancaster, 1982; Tsoar, 1982; Rubin & Hunter, 1985; Glennie, 1987; Rubin, 1990). According to Rubin & Hunter (1987), the trend of aeolian bedforms is determined by the maximum gross bedform-normal rule. Longitudinal bedforms can form in bidirectional wind regimes if the divergence angle between dominant and subordinate transport (winds) is larger than 90–112°. Based on the orientation of the draa ridges, it is inferred that the draas formed in an overall bidirectional wind regime with winds alternating from northern and southeastern directions. As draas are large bedforms with a long reconstitution time (10^3–10^5 yrs, cf. Lancaster, 1988), they respond only to long-term changes in wind regime. In the present case there is evidence of an early phase of lateral draa migration followed by a second phase of vertical accretion possibly coupled with a long-term change in the balance between the alternating winds, and thereby in the direction of the resultant transport (Clemmensen, 1989).

The superimposed longitudinal dunes seem to have existed in equilibrium with the larger draas, but the dunes responded to a different scale of airflow variability (cf. Lancaster, 1988). Thus it is inferred that the superimposed dunes were controlled by seasonal wind regimes. Because of the longitudinal nature of the dunes it is safe to conclude that they were controlled by primary winds and not deflected secondary airflow (cf. Tsoar, 1982). The trends of the superimposed longitudinal dunes were 55–235° as deduced from outcrop data, especially the orientation of the inter-digitation zones. Seasonal wind shifts from northwest or north to southeast or east are therefore deduced. The exact direction of these seasonal winds cannot be reconstructed, but the transport ratio must have been near unity, which has also been deduced from studies of modern longitudinal dunes (Tsoar, 1982).

The Southern Permian Basin was situated at *c*.10° northern palaeolatitude during the deposition of the Yellow Sands, or in an ancient tropical–subtropical desert belt (cf. Glennie, 1983). It seems reasonable therefore to interpret the northwest or north winds as winter season trade winds and the southeast or east winds as summer 'monsoon' winds. These 'monsoon' winds seem to have been relatively dry. In contrast to this view, Glennie (1982, 1983) has considered the Yellow Sands to have formed during the influence of strong unidirectional winds from the northeast, and Sneh (1988) has postulated unidirectional winds from the north during the formation of the Rotliegend.

SUMMARY AND CONCLUSIONS

1 The Early Permian Yellow Sands form several large parallel-trending sand ridges representing partly preserved linear draas. The draas experienced an early phase of lateral migration (oblique bedforms) and a second phase of vertical accretion (longitudinal bedforms) before they were covered by aeolian sand sheet deposits, and ultimately were preserved beneath low-energy marine deposits.
2 Studies of aeolian architecture and computer-graphics modelling indicate that the draas were superimposed by sinuous longitudinal dunes. The superimposed dunes were reversing, non-migrating and had sinuosities that migrated along-crest. The longitudinal dunes were separated by inter-dune areas in which shallow wind scours formed frequently.
3 In the final phase of draa sedimentation, compound longitudinal bedforms formed in a bimodal wind regime with alternating northwest or north winter winds and southeast or east summer 'monsoon' winds.

ACKNOWLEDGEMENTS

This study was supported by the Geological Institute, University of Copenhagen, and by the Danish Ministry of Energy. We wish to thank M. Tucker and G. Larwood (Durham) for support during the initial phase of the investigation, and the quarry owners for permission to work in the quarries. Journal referees are thanked for constructive reviews.

REFERENCES

Ahlbrandt, T.S. & Fryberger, S.G. (1980) Geological and paleoecologic studies of Nebraska Sand Hills. *US Geol. Surv. Prof. Pap. 1120-A*, 24 pp.

Bell, C.M. (1991) The relationships between sedimentary structures, transport directions and dune types in Mesozoic aeolian sandstones, Atacama Region, Chile. *Sedimentology* **38**, 289–300.

Breed, C.S., Fryberger, J.G., Andrews, S., McCauley, C., Lennartz, F., Gebel, D. & Horstman, K. (1979) Regional studies of sand seas, using Landsat (ERTS) imagery. In: *A Study of Global Sand Seas* (Ed. McKee, E.D.) pp. 305–397. US Geol. Surv. Prof. Pap. 1052.

Chrintz, T. & Clemmensen, L.B. (1989) Compound linear dunes in the Lower Permian Yellow Sands, NE England: aeolian architecture and dynamics. *Abstracts, 28th Int. Geol. Congress, Washington, DC*, 290–291.

Chrintz, T. & Clemmensen, L.B. (1990) Draa reconstruction, the Permian Yellow Sands, northeast England. *Abstracts, 13th Int. Sed. Congress, Nottingham*, 94–95.

CLEMMENSEN, L.B. (1989) Preservation of interdraa and draa plinth deposits by the lateral migration of large linear draas (Lower Permian Yellow Sands, northeast England). *Sed. Geol.* **65**, 139–151.

CLEMMENSEN, L.B. (1991a) *Æolisk arkitektur og aflejrings-dynamik. En undersøgelse af ni æoliske aflejringssystemer (Prækambrium–Recent) fra den nordatlantiske region.* Unpublished Dr Scient. thesis, University of Copenhagen.

CLEMMENSEN, L.B. (1991b) Aeolian architecture and reservoir characteristics. *Abstracts, Characterisation of Fluvial and Aeolian Reservoirs, Aberdeen, 24–27 March.*

GLENNIE, K.W. (1970) Desert sedimentary environments. In: *Developments in Sedimentology 24*. Elsevier, Amsterdam, 222 pp.

GLENNIE, K.W. (1982) Early Permian (Rotliegendes) palaeowinds of the North Sea. *Sed. Geol.* **34**, 245–265.

GLENNIE, K.W. (1983) Lower Permian Rotliegend desert sedimentation in the North Sea area. In: *Aeolian Sediments and Processes*. (Ed. Brookfield, M.E. & Ahlbrandt, T.S.) pp. 521–541. Developments in Sedimentology 38. Elsevier, Amsterdam.

GLENNIE, K.W. (1987) Desert sedimentary environments, present and past—a summary. *Sed. Geol.* **50**, 135–165.

GLENNIE, K.W. & BULLER, A.T. (1983) The Permian Weissliegend of NW Europe: the partial deformation of aeolian dune sands caused by the Zechstein transgression. *Sed. Geol.* **35**, 43–81.

HUNTER, R.E. (1981) Stratification styles in aeolian sandstones: some Pennsylvanian to Jurassic examples from the Western Interior, USA. In: *Recent and Ancient Nonmarine Depositional Environments* (Ed. Etheridge, F.G. & Flores, R.M.) pp. 315–329. Soc. Econ. Paleontol. Mineral. Spec. Pub. 31. SEPM, Tulsa.

KOCUREK, G., TOWNSLEY, M., YEH, E., SWEET, M. & HAVHOLM, K. (1992) Dune and dune-field development on Padre Island, Texas: implications for interdune deposition and water-table controlled accumulation. *J. Sed. Petrol.* **62**, 622–635.

LANCASTER, N. (1982) Linear dunes. *Prog. Phys. Geog.* **6**, 475–504.

LANCASTER, N. (1983) Linear dunes of the Namib sand sea. *Z. Geomorph. NF* **45**, 27–49.

LANCASTER, N. (1988) The development of large aeolian bedforms. *Sed. Geol.* **55**, 69–89.

MADER, D. & YARDLEY, M.J. (1985) Migration, modification and merging in aeolian systems and the significance of the depositional mechanisms in Permian and Triassic dune sands of Europe and North America. *Sed. Geol.* **43**, 85–218.

MCKEE, E.D. (1979a) Sedimentary structures in dunes. In: *A Study of Global Sand Seas* (Ed. McKee, E.D.) pp. 83–134. US Geol. Surv. Prof. Pap. 1052.

MCKEE, E.D. (1979b) Ancient sandstones considered to be aeolian. In: *A Study of Global Sand Seas* (Ed. McKee, E.D.) pp. 187–238. US Geol. Surv. Prof. Pap. 1052.

PRYOR, W.A. (1971) Petrology of the Permian Yellow Sands of northeastern England and their North Sea equivalents. *Sed. Geol.* **6**, 221–254.

RUBIN, D.M. (1987) Cross-bedding, bedforms and paleocurrents. Concepts in sedimentology and paleontology 1. *Soc. Econ. Paleontol. Mineral.* 1–187.

RUBIN, D.M. (1990) Lateral migration of linear dunes in the Strzelcki desert, Australia. *Earth Surf. Proc. Landf.* **15**, 1–14.

RUBIN, D.M. & HUNTER, R.E. (1985) Why deposits of longitudinal dunes are rarely recognized in the geological record. *Sedimentology* **32**, 147–157.

RUBIN, D.M. & HUNTER, R.E. (1987) Bedform alignment in directionally varying flows. *Science* **237**, 276–278.

SMITH, D.B. & FRANCIS, E.A. (1967) Geology of the country between Durham and West Hartlepool. Mem. Geol. Surv. GB 27, 354 pp.

SNEH, A.K. (1988) Permian dune patterns in Northwestern Europe challenged. *J. Sed. Petrol.* **58**, 44–51.

STEELE, R.P. (1983) Longitudinal draa in the Permian Yellow Sands of north-east England. In: *Eolian Sediments and Processes* (Ed. Brookfield, M.E. & Ahlbrandt, T.S.) pp. 543–550. Developments in Sedimentology 38. Elsevier, Amsterdam.

TSOAR, H. (1982) Internal structure and surface geometry of longitudinal (seif) dunes. *J. Sed. Petrol.* **52**, 823–831.

YARDLEY, M.J. (1984) Cross-bedding in the Permian Yellow Sands of County Durham. *Proc. Yorkshire Geol. Soc.* **45**, 11–18.

Index

Page numbers in *italic* refer to figures and tables.